Guangxi Academy of Agricultural Sciences
2019 Sugarcane Research and Development Report

广西农业科学院
甘蔗发展报告2019

农业农村部广西甘蔗生物技术与遗传改良重点实验室
广西甘蔗遗传改良重点实验室

Key Laboratory of Sugarcane Biotechnology and Genetic Improvement (Guangxi), Ministry of Agriculture and Rural Affairs, P.R. China
Guangxi Key Laboratory of Sugarcane Genetic Improvement

中国农业出版社
北　京

Guangxi Academy of Agricultural Sciences
2019 Sugarcane Research and Development Report

主　　编：李杨瑞　黄东亮　陈忠良
编著人员：吴建明　韦国才　Prakash Lakshmanan
　　　　　王维赞　邓智年　刘昔辉　范业赓　宋修鹏
　　　　　黄　杏　莫璋红　徐　林　陈荣发　林　丽
　　　　　廖　芬　汪　淼　吴凯朝　秦翠鲜　丘立杭
　　　　　桂意云　翁梦苓　李傲梅　刘　璐　陈　莉
　　　　　莫善平

Chief editors: Yang-rui Li, Dong-liang Huang, Zhong-liang Chen
Associated editors:
　　　　Jian-ming Wu, Guo-cai Wei, Prakash Lakshmanan,
　　　　Wei-zan Wang, Zhi-nian Deng, Xi-hui Liu, Ye-geng Fan,
　　　　Xiu-peng Song, Xing Huang, Zhang-hong Mo, Lin Xu,
　　　　Rong-fa Chen, Li Lin, Fen Liao, Miao Wang, Kai-chao Wu,
　　　　Cui-xian Qin, Li-hang Qiu, Yi-yun Gui, Meng-ling Weng,
　　　　Ao-mei Li, Lu Liu, Li Chen, Shan-ping Mo

Contents 目 录

领导重视 IMPORTANT GUESTS / 1

1 承担科研项目 RESEARCH PROJECTS / 4

    1.1　2019年省部级以上项目 Projects Approved in 2019 / 4

    1.2　2019年承担其他委托项目 Other Projects Ongoing in 2019 / 15

2 成果 ACHIEVEMENTS / 20

    2.1　获奖成果 Awards / 20

    2.2　专利 Patents / 20

        2.2.1　申请专利 Application of Patents / 20

        2.2.2　授权专利 Authorized Patents / 27

    2.3　软件著作权 Software Copyright / 34

    2.4　标准 Standard / 34

3 发表论著 PUBLICATIONS / 37

    3.1　发表期刊论文 Journal Papers / 37

    3.2　发表国际会议论文 International Conference Papers / 45

4 合作与交流 COOPERATION AND EXCHANGE / 47

    4.1　实验室人员参加国内外学术交流记录
　　　Important Academic Exchange Activities / 47

        4.1.1　参加国外学术交流记录 International Academic Exchange Activities / 47

        4.1.2　参加国内学术交流记录 Domestic Academic Exchange Activities / 51

    4.2　国内外专家来实验室进行学术交流 Academic Exchange
　　　Activities with Foreign and Domestic Visitors in Laboratory / 57

5 甘蔗科研进展 PROGRESS IN SUGARCANE RESEARCH / 61

    5.1　甘蔗种质创新与育种
　　　Sugarcane Germplasm Innovation and Breeding / 61

        5.1.1 甘蔗种质创新 Sugarcane Germplasm Innovation / 61
        5.1.2 甘蔗高效育种 Efficient Sugarcane Breeding / 70
    5.2 甘蔗栽培及生理 Sugarcane Cultivation and Physiology / 72
        5.2.1 甘蔗生理生态研究 Study on Physiology and Ecology of Sugarcane / 72
        5.2.2 甘蔗轻简栽培技术 Simplified Cultivation Technology for Sugarcane / 89
        5.2.3 甘蔗机械化研究 Sugarcane Mechanization Research / 99
    5.3 甘蔗功能基因组学研究 Functional Genomics of Sugarcane / 101
        5.3.1 组学研究 Genomic / 101
        5.3.2 基因克隆 Gene Clone / 106
        5.3.3 转基因研究 Genetic Modification / 113
    5.4 甘蔗病虫害致病机理及生物防治 Pathogenesis Mechanism and Biological Control of Sugarcane Diseases and Pests / 114
        5.4.1 甘蔗虫害生物防治 Biological Control of Sugarcane Pests / 114
        5.4.2 甘蔗病害防治及机理研究 Sugarcane Disease Control and Mechanism Research / 119
    5.5 甘蔗生物固氮机理及氮高效利用 Mechanism of Biological Nitrogen Fixation and Efficient Nitrogen Utilization in Sugarcane / 122
        5.5.1 甘蔗生物固氮机理 Mechanism of Biological Nitrogen Fixation in Sugarcane / 122
        5.5.2 甘蔗氮高效利用 Efficient Nitrogen Utilization in Sugarcane / 123

6 附录 APPENDIX / 132

    6.1 实验室学术委员会和固定人员组成 Academic Committee and Staff / 132
    6.2 博士后培养和研究生教育 Postdoctoral Fellow Training and Postgraduate Education / 138

领导重视
IMPORTANT GUESTS

　　2019年10月20日，农业农村部科技教育司司长廖西元、中国工程院院士万建民参观考察了甘蔗新品种展示区。

　　Xi-yuan Liao, Director-general of the Department of Science, Technology and Education, Ministry of Agricultural and Rural Affairs, and Jian-min Wan, Academician of the Chinese Academy of Engineering visited the exhibition field of new sugarcane varieties at Guangxi Academy of Agricultural Sciences.

2019年7月24日，广西壮族自治区农业农村厅农机化管理处处长陈锡诗一行3人到广西农业科学院甘蔗研究所甘蔗农机农艺融合示范基地调研指导。

Xi-shi Chen, director of the Division of the Agricultural Mechanization Management, Guangxi Department of Agricultural and Rural Affairs, visited the demonstration base on integrated technology of sugarcane agricultural machinery and agronomy on July 24, 2019.

2019年4月26日，广西壮族自治区糖业发展办公室刘全跃副巡视员到广西农业科学院甘蔗研究所作"广西糖业及甘蔗种业改革与发展的思考"专题报告。

Quan-yue Liu, deputy director and deputy counsel of the office of Guangxi Sugar Industry Development Office gave a presentation titled "Reflections on the Reform and Development of Guangxi Sugar and Sugarcane Seed Industry" on April 26, 2019.

2019年1月30日，广西壮族自治区科技厅农村处副处长韦昌联到广西农业科学院甘蔗研究所合作良种繁育基地考察。

Chang-lian Wei, deputy director of the Division of Rural Affairs, Guangxi Department of Science and Technology, visited the cooperative propagation base of new sugarcane varieties on January 30, 2019.

1 承担科研项目
RESEARCH PROJECTS

2019年共承担广西壮族自治区级以上在研项目共54项（表1、表2），其中新立项目11项，资助科研经费共计1 164万元，包括国家重点研发计划课题1项，立项经费399万元；国家自然科学基金项目6项，立项经费共197万元；广西科技计划项目2项，立项经费共520万元；广西自然科学基金项目2项，立项经费共48万元（表1）；同年新增广西农业科学院院基金和基本科研业务专项9项，立项经费81万元（表3）。

In 2019, there were total 54 ongoing projects from provincial or national departments (Tables 1 and 2). Among them, 11 provincial or national projects were newly approved with a total fund of 11.64 million yuan, One project was funded by National Key Research and Development Program of China with a budget of 3.99 million yuan, 6 were supported by National Natural Science Foundation of China with a total budget of 1.97 million yuan, 2 were from Guangxi Government R & D Program with a budget of 5.2 million yuan, and 2 were funded by Guangxi Natural Science Foundation with a budget of 0.48 million yuan (Table 1). Guangxi Academy of Agricultural Science had funded 9 projects with a total budget of 0.81 million yuan through its Fundamental Research Fund(Table 3).

1.1 2019年省部级以上项目 Projects Approved in 2019

2019年省部级以上项目获立项11项，累计经费1 194万元，其中当年下拨经费776.6万元（表1）。

In 2019, there were 11 projects approved by provincial or national government with a total fund of 11.94 million yuan, with the current year budget totaling 7.766 million yuan (Table1).

表1 2019年新增省部级以上科研项目一览表

Table 1 Provincial and national government funded projects approved in 2019

序号 No.	项目名称 Title	项目来源 Source	合同编号 Contract Number	起止时间 Period	立项经费（万元）Total Fund	当年到位经费（万元）Fund for 2019	主持人 PI
1	甘蔗种质资源精准评价与基因发掘 Accurate evaluation of sugarcane germplasms and associated genes discovery	国家科技部 Ministry of Science and Technology of the People's Republic of China	2019YFD1000503	2019-05—2022-12	399	290.4	黄东亮 Dong-liang Huang
2	甘蔗响应甘蔗鞭黑粉菌侵染的生理和分子基础研究 The research of physiological and molecular basis of sugarcane's response to the infection of Sporisorium scitamineum	国家自然科学基金 National Natural Science Foundation of China	31901594	2020-01—2022-12	24	14.4	宋修鹏 Xiu-peng Song
3	基于GISH的甘蔗与河八王杂交后代染色体遗传分析 Chromosome genetic analysis of hybrids between Saccharum officinarum and Narenga porphyrocoma based on GISH	国家自然科学基金 National Natural Science Foundation of China	31901510	2020-01—2022-12	23	13.8	段维兴 Wei-xing Duan
4	甘蔗MITE转座子鉴定及其核心标记开发与数据库构建 Identification of sugarcane MITE transposon and development of their core molecular markers and database construction	国家自然科学基金 National Natural Science Foundation of China	31960416	2020-01—2023-12	32	19.2	刘俊仙 Jun-xian Liu
5	甘蔗鞭黑粉菌有性配合及致病相关基因SCP的功能研究 Function analysis of SCP gene involved in sexual mating and pathogenicity of Sporisorium scitamineum	国家自然科学基金 National Natural Science Foundation of China	31960521	2020-01—2023-12	39	23.4	颜梅新 Mei-xin Yan
6	甘蔗抗黑穗病主效QTLs紧密连锁标记的开发及其辅助育种 Development of markers closely linked to major QTLs for sugarcane smut resistance and marker-assisted selection	国家自然科学基金 National Natural Science Foundation of China	31960450	2020-01—2023-12	39	23.4	高铁静 Yi-jing Gao

(续)

序号 No.	项目名称 Title	项目来源 Source	合同编号 Contract Number	起止时间 Period	立项经费（万元）Total Fund	当年到位经费（万元）Fund for 2019	主持人 PI
7	miRNAs在乙烯利诱导甘蔗糖分积累中的调控作用研究 The role of miRNAs in regulating sugar accumulation induced by ethephon in sugarcane	国家自然科学基金 National Natural Science Foundation of China	31960449	2020-01—2023-12	40	24	汪淼 Miao Wang
8	广西甘蔗遗传改良重点实验室 Guangxi Key Laboratory of Sugarcane Genetic Improvement	广西科技计划项目 Guangxi R & D Program	—	2019-01—2020-12	120	120	李杨瑞 Yang-rui Li
9	来宾市国家农业科技园区创新基础能力建设 Construction of basic innovation ability of National Agricultural Science and Technology Park of China (Laibin city)	广西科技计划项目 Guangxi R & D Program	桂科 AD19245080	2020-01—2022-12	400	200	丘立杭 Li-hang Qiu
10	产ABA慢生根瘤菌增强甘蔗耐瘠抗旱能力的机理研究 Analysis of physiological and molecular changes in sugarcane to improve the tolerance of infertility and drought stress by ABA-producing Bradyrhizobium inoculation	广西自然科学基金 Guangxi Natural Science Foundation	2019GXNSFDA185004	2019-09—2023-09	40	40	李长宁 Chang-ning LI
11	机收碾压对蔗地土壤和甘蔗生理生化活性的影响及其与宿根性强弱的关系研究 The effect of rolling on soil, physiology and biochemistry of sugarcane harvested by machine and their relationship with ratooning ability	广西自然科学基金 Guangxi Natural Science Foundation	2019GXNSFBA245005	2020-01—2023-01	8	8	邓宇驰 Yu-chi Deng
合计	11项	—	—	—	1164	776.6	—

2019年省部级以上项目在研项目共43项（表2）。

In 2019, 43 projects from provincial or national departments were undertaking (Table 2).

表2　2019年自治区级以上在研项目一览表

Table 2　Ongoing projects funded by provincial or national government departments in 2019

序号 No.	项目名称 Title	项目来源 Source	合同编号 Contract Number	起止时间 Period	立项经费（万元）Total Fund	当年到位经费（万元）Fund for 2019	主持人 PI
1	甘蔗化肥农药减施增效技术集成研究 Research on the integrated Technology of improving benefits by reducing chemical fertilizer and pesticide on sugarcane	国家科技部 Ministry of Science and Technology of the People's Republic of China	2018YFD0201103	2018-07—2020-12	509	96.36	谭宏伟 Hong-wei Tan
2	伯克氏固氮菌GXS16与甘蔗根系高效联合固氮的生理和分子基础研究 Physiological and molecular basis of high nitrogen fixation efficiency of endophytic diazotroph Burkholderia sp GXS16 associated with sugarcane root	国家自然科学基金 National Natural Science Foundation of China	31801288	2019-01—2021-12	24	1.58	李长宁 Chang-ning Li
3	甘蔗抗梢腐病氮代谢和系统获得性抗性途径关键组分γ-谷氨酰转移酶基因 SoGGT1 克隆及功能鉴定 Cloning and function identification of γ-glutamyltransferase gene SoGGT1, important component of nitrogen metabolism and systemic acquired resistance pathways during sugarcane resistant to pokkahboeng disease	国家自然科学基金 National Natural Science Foundation of China	31801422	2019-01—2021-12	23	1.51	王泽平 Ze-ping Wang
4	蔗叶与蔗叶生物炭还田下其C、N协同归还研究 The C, N return of sugarcane trash and biochar	国家自然科学基金 National Natural Science Foundation of China	31860350	2019-01—2022-12	34	2.23	刘昔辉 Xi-hui Liu

(续)

序号 No.	项目名称 Title	项目来源 Source	合同编号 Contract Number	起止时间 Period	立项经费 （万元） Total Fund	当年到位经费 （万元） Fund for 2019	主持人 PI
5	物种品种资源保护费项目 Project on germplasms conservation	广西壮族自治区农业厅 Department of Agriculture and Rural Affairs of Guangxi	—	2018-01— 2018-12	183	0	李杨瑞 Yang-rui Li
6	甘蔗微型反向重复转座元件的系统分离及其开发和应用研究 The systematic isolation, development and application of sugarcane MITEs	广西自然科学基金 Guangxi Natural Science Foundation	2018GXNSFDA294004	2019-01— 2022-12	40	0	刘俊仙 Jun-xian Liu
7	不同品种甘蔗根瘤菌的多样性研究 Study on the diversity of Rhizobium in different sugarcane varieties	广西自然科学基金 Guangxi Natural Science Foundation	2018GXNSFAA281152	2019-01— 2021-12	12	0	林丽 Li Lin
8	甘蔗抗梢腐病氮代谢和系统获得性抗性途径关键组分γ-谷氨酰转移酶基因SoGGT1克隆及功能鉴定 Cloning and function identification of γ-glutamyltransferase gene SoGGT1, important component of nitrogen metabolism and systemic acquired resistance pathways during sugarcane resistant to pokkahboeng disease	广西自然科学基金 Guangxi Natural Science Foundation	2018GXNSFAA281213	2019-01— 2021-12	12	0	王泽平 Ze-ping Wang
9	甘蔗宿根矮化病致病基因pglA功能解析及互作基因的挖掘 Funtional analysis of pathogencity gene pglA in Leifsonia xyli subsp. xyli and excavation of its interaction genes	广西自然科学基金 Guangxi Natural Science Foundation	2018GXNSFAA294041	2019-01— 2020-12	20	10	张小秋 Xiao-qiu Zhang
10	转录组测序揭示GA₃抑制甘蔗分蘖的调控机制 The regulatory mechanism of GA_3 inhibiting tillering in sugarcane revealed by transcriptome sequencing	广西自然科学基金 Guangxi Natural Science Foundation	2018GXNSFAA138149	2018-07— 2021-07	12	0	丘立杭 Li-hang Qiu

(续)

序号 No.	项目名称 Title	项目来源 Source	合同编号 Contract Number	起止时间 Period	立项经费（万元）Total Fund	当年到位经费（万元）Fund for 2019	主持人 PI
11	SofSPS DⅢ家族基因在高糖甘蔗及其后代种质间的遗传分析 Genetic analysis of SofSPS DⅢ family genes in high sugar sugarcane and its progeny germplasm	广西自然科学基金 Guangxi Natural Science Foundation	2018GXNSFAA138013	2018-07—2021-07	10	0	陈忠良 Zhong-liang Chen
12	甘蔗SofSPS DⅢ基因参与甘蔗生长和蔗糖积累的功能鉴定 Functional identification of SofSPS DⅢ gene involved in growth and sucrose accumulation in sugarcane	广西自然科学基金 Guangxi Natural Science Foundation	2018GXNSFAA138013	2018-07—2020-07	20	10	蔡翠鲜 Cui-xian Qin
13	甘蔗宿根矮化病灌病株根际微生物群落结构重建及拮抗菌的鉴定与利用研究 Study on re-construction of soil microbial community structure in rhizosphere of sugarcane infected ratoon stunting disease (RSD) and indentification, utilization of its antagonistic microbes	国家自然科学基金 National Natural Science Foundation of China	31760368	2018-01—2021-12	38	15.2	谭宏伟 Hong-wei Tan
14	斑茅割手密复合体后代宿根性与SSR标记的关联分析 Association analysis between ratoon ability and SSR markers in descendant of intergeneric hybrid complex of Erianthus arundinaceus × Saccharum spontaneum	国家自然科学基金 National Natural Science Foundation of China	31760415	2018-01—2021-12	39	15.6	张保青 Bao-qing Zhang
15	薪菠萝灰粉蚧不同地理种群高温胁迫的生殖差异及生理响应机制 The reproductive differences and physiological response mechanism of different geographic population of Dysmicoccus neobrevipes Beardsley under high temperature stress	国家自然科学基金 National Natural Science Foundation of China	31760540	2018-01—2021-12	38	15.2	覃振强 Zhen-qiang Qin

(续)

序号 No.	项目名称 Title	项目来源 Source	合同编号 Contract Number	起止时间 Period	立项经费（万元）Total Fund	当年到位经费（万元）Fund for 2019	主持人 PI
16	转录组动态揭示GA_3和PP_{333}影响甘蔗分蘖的分子调控机制 Transcriptomic dynamic reveals molecular mechanism of GA_3 and PP_{333} regulating tillering in sugarcane	国家自然科学基金 National Natural Science Foundation of China	31701363	2018-01—2020-12	23	9.2	丘立杭 Li-hang Qiu
17	甘蔗优良新品种选育及推广 Breeding and extension of new elite sugarcane varieties	广西科技计划项目 Guangxi R & D Program	桂科AA17202042	2017-09—2020-12	1 400	400	谭宏伟 Hong-wei Tan
18	高通量甘蔗育种技术体系研发 Research and development of a high throughput sugarcane breeding technology system	广西科技计划项目 Guangxi R & D Program	桂科AA17202012	2017-09—2019-12	2 000	500	黄东亮 Dong-liang Huang
19	赤霉素合成关键组分GA_{20}、DELLA和GID1基因调控甘蔗节间伸长的机制研究 Mechanism of GA_{20}, DELLA and GID1 genes, the key components of gibberellin synthesis, regulating sugarcane internode elongation	广西自然科学基金 Guangxi Natural Science Foundation	2017GXNSFBA198050	2017-09—2020-09	9	0	陈荣发 Rong-fa Chen
20	甘蔗及其近缘属野生种高固氮基因型筛选及高固氮机制研究 Screening of high nitrogen fixation (HNF) genotypes from sugarcane and its wild species and the mechanism of HNF	广西自然科学基金 Guangxi Natural Science Foundation	2017GXNSFAA198029	2017-09—2020-09	12	0	罗霆 Ting Luo
21	桂糖系列甘蔗品种LTR反转座子的分子标记精准鉴定及评价研究 LTR retrotransposons based molecular marker identification and evaluation of Guitang sugarcane varieties	广西自然科学基金 Guangxi Natural Science Foundation	2017GXNSFAA198032	2017-09—2020-09	8	0	刘俊仙 Jun-xian Liu

(续)

序号 No.	项目名称 Title	项目来源 Source	合同编号 Contract Number	起止时间 Period	立项经费（万元）Total Fund	当年到位经费（万元）Fund for 2019	主持人 PI
22	ABA调控蔗芽应答低温胁迫的作用机制研究 Mechanisms of ABA on sugarcane buds response to low temperature	国家自然科学基金 National Natural Science Foundation of China	31660356	2017-01—2020-12	39	2.7	黄杏 Xing huang
23	甘蔗螟虫卵寄生蜂种群动态及其关键影响因子研究 Studies on the population dynamic of egg parasitoids for sugarcane borers and its key factors	国家自然科学基金 National Natural Science Foundation of China	31660534	2017-01—2020-12	40	2.75	潘雪红 Xue-hong Pan
24	基于RNA-Seq技术解析甘蔗分蘖习性的激素调控机理 Analysis of the mechanism of phytohormone regulating tillering in sugarcane based on RNA-Seq technology	广西自然科学基金 Guangxi Natural Science Foundation	2016GXNSFBA380034	2016-09—2019-08	5	0	丘立杭 Li-hang Qiu
25	螟黑卵蜂生物学特性、发生动态及对蔗螟自然控制效能研究 Studies on the biological characteristics, population dynamics of Telenomus and its natural control efficiency against sugarcane borer	广西自然科学基金 Guangxi Nature Science Foundation	2016GXNSFBA380125	2016-09—2019-08	5	0	潘雪红 Xue-hong Pan
26	割手密分蘖QTL利用的分子标记辅助选择及遗传分析 Using QTL of tiller for molecular marker assisted selection and genetic analysis in Saccharum spontaneum L.	广西自然科学基金 Guangxi Nature Science Foundation	2016GXNSFBA380225	2016-09—2019-08	5	0	杨翠芳 Cui-fang Yang
27	氮高效甘蔗遗传多样性及其产量性状的关联分析 Genetic diversity and association analysis of nitrogen use efficiency in surgarcane	广西自然科学基金 Guangxi Nature Science Foundation	2016GXNSFBA380138	2016-09—2019-08	5	0	李翔 Xiang Li

(续)

序号 No.	项目名称 Title	项目来源 Source	合同编号 Contract Number	起止时间 Period	立项经费（万元）Total Fund	当年到位经费（万元）Fund for 2019	主持人 PI
28	甘蔗NBS-LRR类抗梢腐病基因的定量表达和功能分析 The quantity expression and function analyse of NBS-LRR genes against pokkah boeng disease in sugarcane	广西自然科学基金 Guangxi Nature Science Foundation	2016GXNSFBA380046	2016-09—2019-08	5	0	王泽平 Ze-ping Wang
29	甘蔗蔗糖合成酶基因（ScSuSy4）功能分析 Function of sucrose synthase gene (ScSuSy4) from sugarcane	广西自然科学基金 Guangxi Nature Science Foundation	2016GXNSFBA380168	2016-09—2019-08	5	0	桂意云 Yi-yun Gui
30	机械收获条件下宿根甘蔗根系生长发育及理化特性的研究 Study on growing development and physicochemical properties of ratoon cane roots under the condition of mechanical harvesting	广西自然科学基金 Guangxi Nature Science Foundation	2016GXNSFBA380206	2016-09—2019-08	5	0	李毅杰 Yi-jie Li
31	内生固氮菌与甘蔗根系互作的代谢多样性分析 Metabolite diversity analysis in sugarcane roots inoculated with endophytic diazotroph	广西自然科学基金 Guangxi Nature Science Foundation	2016GXNSFAA380126	2016-09—2019-08	5	0	李长宁 Chang-ning Li
32	基于GBS技术的甘蔗SNP开发及其在图谱构建和遗传分析中的应用 Development of SNP markers using Genotyping-By-Sequencing (GBS) for mapping and genetic analysis in sugarcane	广西自然科学基金 Guangxi Nature Science Foundation	2016GXNSFAA380010	2016-09—2019-08	5	0	高轶静 Yi-jing Gao
33	氮化肥减量有机培肥对甘蔗氮代谢及产量、品质的影响 Effects on nitrogenous metabolism, yeild and quality of sugarcane by reducing nitrogen fertilization while adding organic fertilizer	广西自然科学基金 Guangxi Nature Science Foundation	2016GXNSFAA380020	2016-09—2019-08	5	0	谢金兰 Jin-lan Xie

(续)

序号 No.	项目名称 Title	项目来源 Source	合同编号 Contract Number	起止时间 Period	立项经费 （万元） Total Fund	当年到位经费 （万元） Fund for 2019	主持人 PI
34	合河八王血缘的甘蔗抗病新种质创制及其分子细胞遗传学基础研究 Studies on the establishment and molecular cytogenetics of new sugarcane germplasm resistance to diseases with *Narenga* relative	广西科技计划项目 Guangxi R & D Program	桂科 AB16380157	2016-09— 2019-12	80	0	段维兴 Wei-xing Duan
35	斑割复合体利用适宜回交世代分子遗传学基础及多抗优良亲本创新 Analysis of the molecular genetics of the appropriate backcross progenis of *arundinaceus-spontaneun* complex and creation of elite breeding parents	广西科技计划项目 Guangxi R & D Program	桂科 AB16380126	2016-09— 2019-12	80	0	张革民 Ge-min Zhang
36	适合机收的"双高"甘蔗新品种高效栽培关键技术研究与示范 Studies on key cultivation technologies of high benefits and their demonstrations with the "double - high" new sugarcane varieties suitable for mechanical harvest	广西科技计划项目 Guangxi R & D Program	桂科 AB16380177	2016-09— 2019-12	80	0	王伦旺 Lun-wang Wang
37	广西丘陵地带甘蔗精量灌溉关键技术研究、集成示范及推广应用 Researches on the key technologies of sugarcane precision irrigation and their integrated demonstration and application in hilly region of Guangxi province	广西科技计划项目 Guangxi R & D Program	桂科 AB16380258	2016-09— 2019-12	150	0	李鸣 Ming Li
38	国家糖料产业技术体系建设 Construction of national sugarcane industry and technology system	国家农业农村部 Ministry of Agriculture and Rural Affairs	CARS-170206	2017-01— 2020-12	280	70	谭宏伟 Hong-wei Tan

(续)

序号 No.	项目名称 Title	项目来源 Source	合同编号 Contract Number	起止时间 Period	立项经费（万元）Total Fund	当年到位经费（万元）Fund for 2019	主持人 PI
39	国家糖料产业技术体系建设 Construction of national sugarcane industry and technology system	国家农业农村部 Ministry of Agriculture and Rural Affairs	CARS-170105	2017-01—2020-12	280	70	杨荣仲 Rong-zhong Yang
40	国家糖料产业技术体系建设 Construction of national sugarcane industry and technology system	国家农业农村部 Ministry of Agriculture and Rural Affairs	CARS-170305	2017-01—2020-12	280	70	黄诚华 Cheng-hua Huang
41	国家现代农业产业技术体系广西甘蔗创新团队建设 Construction of Guangxi sugarcane innovation team of national technology system for modern agricultural industry	广西壮族自治区人民政府 The People's Government of Guangxi Zhuang Autonomous Region	nycytxgxcxtd-03-01	2016-01—2020-12	200	40	李杨瑞 Yang-rui Li
42	国家现代农业产业技术体系广西甘蔗创新团队建设 Construction of Guangxi sugarcane innovation team of national technology system for modern agricultural industry	广西壮族自治区人民政府 The People's Government of Guangxi Zhuang Autonomous Region	nycytxgxcxtd-03-02	2016-01—2020-12	125	25	谭宏伟 Hong-wei Tan
43	国家现代农业产业技术体系广西甘蔗创新团队建设 Construction of Guangxi sugarcane innovation team of national technology system for modern agricultural industry	广西壮族自治区人民政府 The People's Government of Guangxi Zhuang Autonomous Region	nycytxgxcxtd-03-03	2016-01—2020-12	125	25	覃振强 Zhen-qiang Qin
合计	43项	—	—	—	6 275	1 382.33	—

1.2 2019年承担其他委托项目 Other Projects Ongoing in 2019

2019年获立项其他委托项目9项，累计经费81万元，目当年到位经费51万元（表3）。

In 2019, 9 projects of the Fundamental Research Fund of Guangxi Academy of Agricultural Science were approved, with the funds of 0.81 million yuan, and the fund in place was 0.51 million yuan in the current year (Table3) .

表3 2019年新增承担其他委托科研项目一览表

Table 3 Projects funded by Guangxi Academy of Agricultural Sciences in 2019

序号 No.	项目名称 Title	项目来源 Source	合同编号 Contract Number	起止时间 Period	立项经费（万元）Total Fund	当年到位经费（万元）Fund for 2019	主持人 PI
1	河八王抗旱/病关键基因挖掘与功能分析 The key drought/disease-tolerant genes mining and their functional analysis for *Narenga porphyrocoma*	广西农业科学院 Guangxi academy of agricultural science	桂农科M19	2019-05—2021-04	10	5	刘昔辉 Xi-hui Liu
2	生物炭对甘蔗土壤氮及氮代谢酶活性的影响 Effects of biochar on nitrogen in soil and nitrogen metabolism enzyme activities in sugarcane	广西农业科学院 Guangxi academy of agricultural science	桂农科M20	2019-05—2021-04	10	5	廖芬 Feng Liao
3	甘蔗防御梢腐病菌侵染生理生化机制 Physiological and biochemical defense mechanisms for sugarcane infected by *Pokkah Boeng*	广西农业科学院 Guangxi academy of agricultural science	桂农科M21	2019-05—2021-04	10	5	王泽平 Ze-ping Wang
4	甘蔗试管苗光合自养生根技术操作规程 Regulations for technical operations of photosynthetic autotrophic rooting technology for sugarcane test-tube seedlings	广西农业科学院 Guangxi academy of agricultural science	桂农科M32	2019-05—2021-04	1	1	何为中 Wei-zhong He
5	伯克氏固氮菌GXS16与甘蔗根系高效联合固氮的生理和分子基础研究 Physiological and molecular basis of high nitrogen fixation efficiency of endophytic diazotroph *Burkholderia* sp GXS16 associated with sugarcane root.	广西农业科学院 Guangxi academy of agricultural science	31801288	2019-05—2021-04	10	5	李长宁 Chang-ning Li

(续)

序号 No.	项目名称 Title	项目来源 Source	合同编号 Contract Number	起止时间 Period	立项经费（万元）Total Fund	当年到位经费（万元）Fund for 2019	主持人 PI
6	甘蔗抗梢腐病氮代谢和系统获得性抗性途径关键组分 γ 谷氨酰转移酶基因 SoGGT1 克隆及功能鉴定 Cloning and function identification of γ-glutamyltransferase gene SoGGT1, important component of nitrogen metabolism and systemic acquired resistance pathways during sugarcane resistant to pokkahboeng disease	广西农业科学院 Guangxi academy of agricultural science	31801422	2019-05—2021-04	10	5	王泽平 Ze-ping Wang
7	蔗叶与蔗叶生物炭还田下其 C、N 协同归还研究 The C, N return of sugarcane trash and biochar	广西农业科学院 Guangxi academy of agricultural science	31860350	2019-05—2021-04	10	5	刘昔辉 Xi-hui Liu
8	甘蔗育种体系创新与新品种选育 Innovation of sugarcane breeding system and its breeding of new varieties	广西农业科学院 Guangxi academy of agricultural science	桂农科Z04	2019-01—2020-12	20	10	吴建明 Jian-ming Wu
9	甘蔗基因组大数据分析与育种创新 Big data analysis of sugarcane genome and its innovation in sugarcane breeding	广西农业科学院 Guangxi academy of agricultural science	桂农科Z07	2019-01—2020-12	20	10	黄东亮 Dong-liang Huang
合计	9项	—	—	—	81	51	—

2019年省部级以下项目在研项目共20项（表4）。

In 2019, 20 projects from Guangxi Academy of Agricultural Sciences were undertaking (Table 4).

表4 2019年承担其他委托在研项目一览表

Table 4 Ongoing projects funded by Guangxi Academy of Agricultural Sciences in 2019

序号 No.	项目名称 Title	项目来源 Source	合同编号 Contract Number	起止时间 Period	立项经费（万元） Total Fund	当年到位经费（万元） Fund for 2019	主持人 PI
1	高产、高糖、抗逆强甘蔗新品种选育 Breeding of new sugarcane varieties with high yield, high sugar and stress resistance	广西农业科学院 Guangxi academy of agricultural science	2015YT01	2015-01—2020-12	180	15	杨荣仲 Rong-zhong Yang
2	甘蔗简化栽培新技术研究 Research on the new simplified cultural techniques for sugarcane	广西农业科学院 Guangxi academy of agricultural science	2015YT02	2015-01—2020-12	120	15	李杨瑞 Yang-rui Li
3	甘蔗功能基因组研究 Research on functional genomics in sugarcane	广西农业科学院 Guangxi academy of agricultural science	2015YT03	2015-01—2020-12	120	10	黄东亮 Dong-liang Huang
4	甘蔗优异野生基因资源发掘利用 Discovery and their utilization of elite genes in the wild germplasm resources of sugarcane	广西农业科学院 Guangxi academy of agricultural science	2015YT04	2015-01—2020-12	120	20	张革民 Ge-min Zhang
5	甘蔗农机农艺融合关键技术集成研究与示范 Research and demonstration of the key technology combined with agricultural machinery and agronomy	广西农业科学院 Guangxi academy of agricultural science	2015YT05	2015-01—2020-12	60	10	王维赞 Wei-zan Wang
6	甘蔗螟虫及主要病害综防体系关键技术的研究 Study on the key techniques of the comprehensive control systems for borers and main diseases in sugarcane	广西农业科学院 Guangxi academy of agricultural science	2015YT06	2015-01—2020-12	60	10	黄诚华 Cheng-hua Huang
7	甘蔗健康种苗技术创新研究 Research on technology innovation of the sugarcane pathogen-free seedlings	广西农业科学院 Guangxi academy of agricultural science	2015YT07	2015-01—2020-12	60	10	李松 Song Li

(续)

序号 No.	项目名称 Title	项目来源 Source	合同编号 Contract Number	起止时间 Period	立项经费（万元）Total Fund	当年到位经费（万元）Fund for 2019	主持人 PI
8	甘蔗生理生化调控技术研究 Research on physiological and biochemical regulation technology in sugarcane	广西农业科学院 Guangxi academy of agricultural science	桂农科2018YT01	2018-01—2020-12	30	10	吴建明 Jian-ming Wu
9	甘蔗抗逆育种与生物学研究 Study on tolerance breeding and its biology in sugarcane	广西农业科学院 Guangxi academy of agricultural science	桂农科2018YT02	2018-01—2020-12	30	10	刘昔辉 Xi-hui Liu
10	甘蔗生物固氮研究 Study on biological Nitrogen fixation in sugarcane	广西农业科学院 Guangxi academy of agricultural science	桂农科2018YT03	2018-01—2020-12	30	10	林丽 Li Lin
11	甘蔗抗病育种研究 Research on disease resistant breeding in sugarcane	广西农业科学院 Guangxi academy of agricultural science	桂农科2018YT04	2018-01—2020-12	30	10	林善海 Shan-hai Lin
12	甘蔗育种及新快繁技术 Breeding and its new rapid propagation techniques in sugarcane	广西农业科学院 Guangxi academy of agricultural science	桂农科2018YT05	2018-01—2020-12	30	10	何为中 Wei-zhong He
13	青年拔尖人才资助项目 Project of top young talent improvement	广西农业科学院 Guangxi academy of agricultural science	桂农科2018YM01	2018-01—2019-12	10	6	刘昔辉 Xi-hui Liu
14	青年拔尖人才资助项目 Project of top young talent improvement	广西农业科学院 Guangxi academy of agricultural science	桂农科2018YM02	2018-01—2019-12	10	6	吴建明 Jian-ming Wu
15	构建基于GIS广西土壤环境质量检测评价信息系统的关键技术研究 Research on key technology construction of the evaluation information system based on GIS for soil environmental quality in Guangxi Province	广西农业科学院 Guangxi academy of agricultural science	桂农科2018ZJ13	2018-01—2018-12	10	10	谭宏伟 Hong-wei Tan

(续)

序号 No.	项目名称 Title	项目来源 Source	合同编号 Contract Number	起止时间 Period	立项经费 (万元) Total Fund	当年到位经费 (万元) Fund for 2019	主持人 PI
16	转录组动态揭示GA$_3$和PP$_{333}$影响甘蔗分蘖的分子调控机制 The molecular mechanism of GA$_3$ and PP$_{333}$ regulating tillering in sugarcane revealed by transcriptomic dynamic	广西农业科学院 Guangxi academy of agricultural science	桂农科2018ZJ14	2018-01—2020-12	10	5	丘立杭 Li-hang Qiu
17	斑茅割手密复合体后代宿根性与SSR标记的关联分析 Association analysis between ratoon ability and SSR markers in descendant of intergeneric hybrid complex of *Erianthus arundinaceus* × *Saccharum spontaneum*	广西农业科学院 Guangxi academy of agricultural science	桂农科2018ZJ16	2018-01—2021-12	10	5	张保青 Bao-qing Zhang
18	新渡萝灰粉蚧不同地理种群高温胁迫的生殖差异及生理响应机制 The reproductive differences and physiological response mechanism of different geographic population of *Dysmicoccus neobrevipes* Beardsley under high temperature stress	广西农业科学院 Guangxi academy of agricultural science	桂农科2018ZJ17	2018-01—2021-12	10	5	覃振强 Zhen-qiang Qin
19	甘蔗宿根矮化病罹病株根际微生物群落结构重建及拮抗菌的鉴定与利用研究 Study on re-construction of soil microbial community structure in rhizosphere of sugarcane infected ratoon stunting disease (RSD) and indentification, utilization of its antagonistic microbes	广西农业科学院 Guangxi academy of agricultural science	桂农科2018ZJ15	2018-01—2021-12	10	5	谭宏伟 Hong-wei Tan
20	甘蔗新品种白条病的抗性评价 Resistant evaluation of *Xanthomonas albilineans* for new sugarcane varieties	广西农业科学院 Guangxi academy of agricultural science	桂农科2018ZJ31	2018-01—2019-12	10	5	魏春燕 Chun-yan Wei
合计	20项	—	—	—	950	187	—

2 成 果
ACHIEVEMENTS

2.1 获奖成果 Awards

2019年获奖成果（表5）。

Awards in 2019 (Table 5).

表5　2019年获奖成果

Table 5　Awards in 2019

序号 No.	获奖成果名称 Title	获奖名称及等级 Award Name and Grade	授奖单位 Granting Organization	获奖人 Winner
1	糖能兼用甘蔗关键技术研究与应用 Research and application of key technologies for sugar and energy dual-purpose sugarcane	广西科技进步奖，二等奖 Guangxi Science and Technology Progress Award (Second Prize)	广西壮族自治区人民政府 The People's Government of Guangxi Zhuang Autonomous Region	李杨瑞，吴建明，邓智年，王伦旺，雷敬超，宋修鹏，杨柳，邢永秀，刘昔辉，张保青，张荣华，唐其展，韦金菊，梁强，李长宁 Yang-rui Li, Jian-ming Wu, Zhi-nian Deng, Lun-wang Wang, Jing-chao Lei, Xiu-peng Song, Liu Yang, Yong-xiu Xing, Xi-hui Liu, Bao-qing Zhang, Rong-hua Zhang, Qi-zhan Tang, Jin-ju Wei, Qiang Liang, Chang-ning Li
2	甘蔗野生种质割手密资源鉴定评价及其抗旱基因挖掘 Evaluation of *Saccharum spontaneum* and discovery of drought-resistant genes	云南省科技进步奖，三等奖 Yunnan Science and Technology Progress Award (Third Prize)	云南省人民政府 The People's Government of Yunnan Province	刘新龙，张革民，刘洋 Xin-long Liu, Ge-min Zhang, Yang Liu

2.2 专利 Patents

2.2.1 申请专利 Application of Patents

2019年申请专利（表6）。

Application of patents in 2019 (Table 6).

表6 2019年申请专利

Table 6 Application of patents in 2019

序号 No.	专利名称 Patent Title	申请时间 Application Time	申请号 Application No.	专利类型 Patent Type	发明人 Inventor
1	一种专用防治甘蔗病虫害的纳米肥及其制备方法 A nano-fertilizer and its preparation method for the prevention and control of sugarcane pests and diseases	2019-08-07	201910724915.X	发明专利 Invention	王泽平，李毅杰，蒋洪涛，梁强，宋修鹏，李长宁，林善海，刘璐，马文清，郭强 Ze-ping Wang, Yi-jie Li, Hong-tao Jiang, Qiang Liang, Xiu-peng Song, Chang-ning Li, Shan-hai Lin, Lu Liu, Wen-qing Ma, Qiang Guo
2	一种甘蔗专用纳米生物有机复合肥及其制备方法 A special nano biome organic compound fertilizer for sugarcane and its preparation method	2019-08-07	201910724909.4	发明专利 Invention	王泽平，刘璐，蒋洪涛，马文清，郭强，李长宁，梁强，李毅杰，宋修鹏，林善海 Ze-ping Wang, Lu Liu, Hong-tao Jiang, Wen-qing Ma, Qiang Guo, Chang-ning Li, Qiang Liang, Yi-jie Li, Xiu-peng Song, Shan-hai Lin
3	一种甘蔗专用纳米水溶性肥及其制备方法 A special nano-water-soluble fertilizer for sugarcane and its preparation method	2019-08-07	201910725121.5	发明专利 Invention	王泽平，林善海，蒋洪涛，梁强，宋修鹏，李毅杰，刘璐，郭强，马文清 Ze-ping Wang, Shan-hai Lin, Hong-tao Jiang, Qiang Liang, Xiu-peng Song, Chang-ning Li, Yi-jie Li, Lu Liu, Qiang Guo, Wen-qing Ma
4	一种甘蔗专用纳米激活增效复混肥及其制备方法 A sugarcane-specific nano-activate-efficiency compound-mixing fertilizer and its preparation method	2019-08-07	201910725123.4	发明专利 Invention	王泽平，宋修鹏，郭强，马文清，刘璐，李长宁，梁强，李毅杰，林善海 Ze-ping Wang, Xiu-peng Song, Qiang Guo, Wen-qing Ma, Hong-tao Jiang, Lu Liu, Chang-ning Li, Qiang Liang, Yi-jie Li, Shan-hai Lin
5	一种甘蔗组培苗甘蔗黑穗病菌接种的方法 A method of inoculation of Sporisorium scitamineum on sugarcane tissue culture seedlings	2019-04-19	201910317934.0	发明专利 Invention	颜梅新，韦金菊，宋修鹏，魏春燕，张小秋 Mei-xin Yan, Jin-ju Wei, Xiu-peng Song, Chun-yan Wei, Xiao-qiu Zhang
6	一种天敌发射器 A device for launching natural enemy of pests	2019-05-13	CN201910394455.9	发明专利 Invention	李德伟，覃振强，罗亚伟，宋修鹏，谭宏伟，何为中，梁阗，丁华珍 De-wei Li, Zhen-qiang Qin, Ya-wei Luo, Xiu-peng Song, Hong-wei Tan, Wei-zhong He, Tian Liang, Hua-zhen Ding

(续)

序号 No.	专利名称 Patent Title	申请时间 Application Time	申请号 Application No.	专利类型 Patent Type	发明人 Inventor
7	一种甘蔗实生苗移栽前梢腐病接种胁迫的方法 A method for inoculation of Fusarium verticillioides before re-planting sugarcane seedlings	2019-09-30	201910943508.8	发明专利 Invention	黄海荣、杨荣仲、徐林、邓宇驰、李翔、经艳、王伦旺、覃仕云、周会、高轶静、谭芳、贤武 Hai-rong Huang, Rong-zhong Yang, Lin Xu, Xiang Li, Yu-chi Deng, Yan Jing, Lun-wang Wang, Shi-yun Tang, Hui Zhou, Yi-jing Gao, Fang Tan, Wu Xian
8	一种甘蔗实生苗单株及组合抗旱筛选方法 A method for screening single and combined sugarcane seedlings with drought-resistance	2019-10-23	201911027414.2	发明专利 Invention	黄海荣、徐林、李翔、邓宇驰、经艳、王伦旺、杨荣仲、周会、高轶静 Hai-rong Huang, Lin Xu, Xiang Li, Yu-chi Deng, Yan Jing, Lun-wang Wang, Rong-zhong Yang, Shi-yun Tang, Hui Zhou, Yi-jing Gao
9	促进甘蔗健康种苗快速生长基质的制备方法及应用 The preparation method and application of the fast-growing matrix for virus-free seedling of sugarcane	2019-08-06	201910722294.1	发明专利 Invention	吴凯朝、王维赞、黄诚梅、徐林、邓智年、庞天、李毅杰、刘晓燕、黄成丰、覃文凭 Kai-chao Wu, Wei-zan Wang, Cheng-mei Huang, Lin Xu, Zhi-nian Deng, Rong-hua Zhang, Tian Pang, Yi-jie Li, Xiao-yan Liu, Cheng-feng Huang, Wen-xian Qin
10	一种甘蔗单芽低温保存方法 A method for low temperature preservation of single buds of sugarcane	2019-07-19	201910654120.6	发明专利 Invention	徐林、吴凯朝、张荣华、邓智年、王维赞、庞天、李毅杰、黄成丰、刘晓燕、黄海荣 Lin Xu, Kai-chao Wu, Rong-hua Zhang, Zhi-nian Deng, Wei-zan Wang, Tian Pang, Yi-jie Li, Xiao-yan Liu, Hai-rong Huang
11	一种甘蔗抗寒组合物 A sugarcane anti-cold composition	2019-04-29	201910352923.6	发明专利 Invention	张保青、邵敏、黄杏、宋修鹏、周忠凤、周珊、杨翠芳、杨丽涛、李杨瑞 Bao-qing Zhang, Min Shao, Xin Huang, Xiu-peng Song, Zhong-feng Zhou, Shan Zhou, Cui-fang Yang, Li-tao Yang, Yang-rui Li
12	一种甘蔗全膜覆盖栽培方法 A cultivation method for sugarcane with whole-film covering	2019-10-28	201910605505.3	发明专利 Invention	陈忠良、黄东亮、秦翠鲜、汪淼、廖芬、范业赓、周丽、谢晓玲 Zhong-liang Chen, Dong-liang Huang, Cui-xian Qin, Miao Wang, Fen Liao, Ye-geng Fan, Li Zhou, Xiao-ling Xie
13	一种甘蔗实生苗假期植梢腐病接种装置 A device of inoculation of Fusarium verticillioides on sugarcane during seedling planting period	2019-07-17	201921121739.2	实用新型 Utility model	黄海荣、林善海、邓宇驰、李翔、经艳、王伦旺、唐仕云、高轶静、徐林、杨荣仲 Hai-rong Huang, Shan-hai Lin, Yu-chi Deng, Xiang Li, Yan Jing, Lun-wang Wang, Shi-yun Tang, Yi-jing Gao, Lin Xu, Rong-zhong Yang

(续)

序号 No.	专利名称 Patent Title	申请时间 Application Time	申请号 Application No.	专利类型 Patent Type	发明人 Inventor
14	一种甘蔗种茎消毒装置 A sugarcane stem disinfection device	2019-07-30	201921208812.X	实用新型 Utility model	黄海荣、邓宇驰、林善海、李翔、经艳、王伦旺、唐仕云、徐林 Hai-rong Huang, Yu-chi Deng, Shan-hai Lin, Xiang Li, Yan Jing, Lun-wang Wang, Shi-yun Tang, Li Xu
15	一种甘蔗种茎切割装置 A sugarcane stem cutting device	2019-07-30	201921209142.3	实用新型 Utility model	黄海荣、李翔、唐仕云、周会、邓宇驰、经艳、王伦旺、徐林 Hai-rong Huang, Xiang Li, Shi-yun Tang, Hui Zhou, Yu-chi Deng, Yan Jing, Lun-wang Wang, Lin Xu
16	一种甘蔗砍收装置 A sugarcane harvesting device	2019-07-30	201921209166.9	实用新型 Utility model	黄海荣、经艳、杨荣仲、李翔、邓宇驰、王伦旺、唐仕云、周会 Hai-rong Huang, Yan Jing, Rong-zhong Yang, Xiang Li, Yu-chi Deng, Lun-wang Wang, Shi-yun Tang, Hui Zhou
17	一种甘蔗田间抗倒伏装置 A field device for protecting sugarcane against lodging	2019-07-30	201921208810.0	实用新型 Utility model	黄海荣、段维兴、周会、邓宇驰、经艳、李翔、唐仕云、王伦旺、谭芳、贤武 Hai-rong Huang, Wei-xing Duan, Hui Zhou, Yu-chi Deng, Yan Jing, Xiang Li, Shi-yun Tang, Lun-wang Wang, Fang Tan, Wu Xian
18	一种甘蔗茎组织切条取样装置 A device for sampling cane stem tissue	2019-06-28	201920993018.4	实用新型 Utility model	陈忠良、黄东亮、秦翠鲜、廖芬、汪淼、周丽、谢晓玲 Zhong-liang Chen, Dong-liang Huang, Cui-xian Qin, Fen Liao, Miao Wang, Li Zhou, Xiao-ling Xie
19	一种甘蔗种物流箱 A box for transporting sugarcane stem	2019-06-26	201920971801.0	实用新型 Utility model	吴凯朝、徐林、张荣华、庞天、李毅杰、刘晓燕、黄成丰、覃文 茺、王维赞、邓智年 Kai-chao Wu, Lin Xu, Rong-hua Zhang, Tian Pang, Yi-jie Li, Xiao-yan Liu, Cheng-feng Huang, Wen-xian Qin, Wei-zan Wang, Zhi-nian Deng
20	一种甘蔗种茎包装装置 A sugarcane stem packaging device	2019-06-26	201920971802.5	实用新型 Utility model	吴凯朝、王维赞、邓智年、徐林、庞天、张荣华、李毅杰、刘晓 燕、黄成丰、覃文茺 Kai-chao Wu, Wei-zan Wang, Zhi-nian Deng, Lin Xu, Rong-hua Zhang, Tian Pang, Yi-jie Li, Xiao-yan Liu, Cheng-feng Huang, Wen-xian Qin

(续)

序号 No.	专利名称 Patent Title	申请时间 Application Time	申请号 Application No.	专利类型 Patent Type	发明人 Inventor
21	一种甘蔗盖膜装置 A sugarcane film-covering device	2019-06-26	201920971175.5	实用新型 Utility model	吴凯朝，王维赞，邓智年，徐林，张荣华，庞天，李毅杰，刘晓燕，黄成丰，覃文芫 Kai-chao Wu, Wei-zan Wang, Zhi-nian Deng, Li Xu, Rong-hua Zhang, Tian Pang, Yi-jie Li, Xiao-yan Liu, Cheng-feng Huang, Wen-xian Qin
22	一种甘蔗单芽包衣装置 A sugarcane single-bud coating device	2019-07-15	201921100884.2	实用新型 Utility model	徐林，张荣华，覃文芫，庞天，李毅杰，黄成丰，吴凯朝，王维赞，邓智年，刘晓燕 Lin Xu, Rong-hua Zhang, Wen-xian Qin, Tian Pang, Yi-jie Li, Cheng-feng Huang, Kai-chao Wu, Wei-zan Wang, Zhi-nian Deng, Xiao-yan Liu
23	一种甘蔗单芽存储用冷库 A refrigeration storage for sugarcane single bud	2019-06-08	201920993599.1	实用新型 Utility model	徐林，吴凯朝，邓智年，张荣华，李毅杰，黄成丰，覃文芫 Lin Xu, Kai-chao Wu, Zhi-nian Deng, Rong-hua Zhang, Yi-jie Li, Xiao-yan Liu, Tian Pang, Wei-zan Wang, Cheng-feng Huang, Wen-xian Qin
24	一种甘蔗单芽补苗装置 A devise for filling gap with sugarcane single-bud seedling	2019-07-15	201921106040.9	实用新型 Utility model	徐林，王维赞，邓智年，吴凯朝，黄成丰，庞天，李毅杰，刘晓燕，张荣华，覃文芫 Lin Xu, Wei-zan Wang, Zhi-nian Deng, Kai-chao Wu, Cheng-feng Huang, Tian Pang, Yi-jie Li, Xiao-Yan Liu, Rong-hua Zhang, Wen-xian Qin
25	一种甘蔗种茎转运装置 A sugarcane stem transport device	2019-06-28	201920994364.4	实用新型 Utility model	邓智年，徐林，吴凯朝，王维赞，张荣华，黄成丰，庞天，李毅杰，刘晓燕，覃文芫 Zhi-nian Deng, Li Xu, Kai-chao Wu, Wei-zan Wang, Rong-hua Zhang, Cheng-feng Huang, Tian Pang, Yi-jie Li, Xiao-yan Liu, Wen-xian Qin
26	一种甘蔗地膜伸缩装置 A sugarcane film extending and retracting device	2019-06-26	201920971153.9	实用新型 Utility model	王维赞，徐林，吴凯朝，邓智年，张荣华，李毅杰，黄成丰，刘晓燕，覃文芫 Wei-zan Wang, Li Xu, Kai-chao Wu, Zhi-nian Deng, Rong-hua Zhang, Tian Pang, Yi-jie Li, Cheng-feng Huang, Xiao-yan Liu, Wen-xian Qin

(续)

序号 No.	专利名称 Patent Title	申请时间 Application Time	申请号 Application No.	专利类型 Patent Type	发明人 Inventor
27	一种甘蔗根部清洗设备 A sugarcane root cleaning device	2019-04-11	201910289511.2	实用新型 Utility model	李翔，李毅杰，雷敬超，梁强，林善海，黄曲燕 Xiang Li, Yi-jie Li, Jing-chao Lei, Qiang Liang, Shan-hai Lin, Qu-yan Huang
28	一种甘蔗根部清洗设备 A sugarcane root cleaning device	2019-04-11	201920486131.3	实用新型 Utility model	李翔，雷敬超，李毅杰，林善海，梁强，黄曲燕 Xiang Li, Jing-chao Lei, Yi-jie Li, Shan-hai Lin, Qiang Liang, Qu-yan Huang
29	一种甘蔗超声波清洗机 A sugarcane ultrasonic cleaner	2019-05-10	201920667134.7	实用新型 Utility model	李翔，梁强，黄曲燕，雷敬超，林善海，李毅杰 Xiang Li, Qiang Liang, Qu-yan Huang, Jing-chao Lei, Shan-hai Lin, Yi-jie Li
30	一种甘蔗根部清洗固定装置 A sugarcane root cleaning fixture	2019-05-10	201920670823.3	实用新型 Utility model	李翔，黄曲燕，雷敬超，李毅杰，梁强，林善海 Xiang Li, Qu-yan Huang, Jing-chao Lei, Yi-jie Li, Qiang Liang, Shan-hai Lin
31	一种甘蔗根部旋转清洗结构 A rotating cleaning structure for the root of sugarcane	2019-05-10	201920670740.4	实用新型 Utility model	李翔，梁强，黄曲燕，雷敬超，林善海，李毅杰 Xiang Li, Qiang Liang, Qu-yan Huang, Jing-chao Lei, Shan-hai Lin, Yi-jie Li
32	一种剑麻水栽培装置 A sisal hemp water cultivation device	2019-06-12	201821574602.8	实用新型 Utility model	覃振强，李德伟，罗亚伟，宋修鹏，魏春燕，张燕杏，韦金菊，张小秋 Zhen-qiang Qin, De-wei Li, Ya-wei Luo, Xiu-peng Song, Chun-yan Wei, Yan-xing Zhang, Jin-ju Wei, Xiao-qiu Zhang
33	一种甘蔗和剑麻育苗专用切根杯 A special cup for cuting root of seedlings of sugarcane and sisal hemp	2019-09-04	201822172793.1	实用新型 Utility model	覃振强，李德伟，宋修鹏，魏春燕，罗亚伟，韦金菊 Zhen-qiang Qin, De-wei Li, Xiu-peng Song, Chun-yan Wei, Ya-wei Luo, Jin-ju Wei
34	一种粉蚧饲养装置 A device for rearing mealybugs (Pseudococcidae)	2019-10-31	201822035398X	实用新型 Utility model	覃振强，李德伟，宋修鹏，罗亚伟，李秀鹏，魏春燕，韦金菊 Zhen-qiang Qin, De-wei Li, Xiu-peng Song, Ya-wei Luo, Chun-yan Wei, Jin-ju Wei
35	一种便捷式甘蔗种植工具 A convenient sugarcane growing tool	2019-10-30	201921844275.8	实用新型 Utility model	邓宇驰，王伦旺，贤武，翁梦苓，刘璐，陈莉 Yu-chi Deng, Lun-wang Wang, Wu Xian, Meng-ling Weng, Lu Liu, Li Chen

(续)

序号 No.	专利名称 Patent Title	申请时间 Application Time	申请号 Application No.	专利类型 Patent Type	发明人 Inventor
36	一种甘蔗补苗工具 A tool for filling gap with sugarcane seeding	2019-10-31	201921860912.0	实用新型 Utility model	邓宇驰、李长宁、宋修鹏、黄海荣、王泽平、李翔、李毅杰、梁强、Yu-chi Deng, Chang-ning Li, Xiu-peng Song, Hai-rong Huang, Ze-ping Wang, Xiang Li, Yi-jie Li, Qiang Liang
37	一种甘蔗实生苗定植工具 A sugarcane seedling planting tool	2019-07-19	201921140019.0	实用新型 Utility model	贤武、王伦旺、邓宇驰、黄海荣、谭芳、黄海荣、经艳、Wu Xian, Lun-wang Wang, Yu-chi Deng, Jiang-xiong Liao, Fang Tan, Hai-rong Huang, Yan Jing
38	一种甘蔗实生苗假植工具 A sugarcane seedling false planting tool	2019-06-30	201920999375.1	实用新型 Utility model	王伦旺、邓宇驰、经艳、谭芳、黄海荣、贤武、唐仕云、周会、Lun-wang Wang, Yu-chi Deng, Yan Jing, Fang Tan, Hai-rong Huang, Wu Xian, Shi-yun Tang, Hui Zhou
39	一种曲面多孔喷头 A curved surface porous nozzle	2019-07-30	201921211700.X	实用新型 Utility model	潘雪红、辛德育 Xue-hong Pan, De-yu Xin
40	一种甘蔗螟虫饲养装置及饲养方法 A device for rearing sugarcane borer and the rearing method	2019-06-18	201910527993.0	实用新型 Utility model	魏吉利、黄诚华、商显坤、潘雪红、林善海 Ji-li Wei, Cheng-hua Huang, Xian-kun Shang, Xue-hong Pan, Shan-hai Lin
41	一种甘蔗螟虫饲养装置 A sugarcane borer rearing device	2019-06-18	201920920556.0	实用新型 Utility model	魏吉利、黄诚华、潘雪红、商显坤、林善海 Ji-li Wei, Cheng-hua Huang, Xue-hong Pan, Xian-kun Shang, Shan-hai Lin
42	一种含有双丙环虫酯和松脂酸钠的超低容量液剂 An ultra-low-volume liquid containing dipropylene cyclobiteest and sodium sodium soslenate	2019-06-28	201910576160.3	实用新型 Utility model	邓宇驰、王伦旺、贤武、吴建明、宋修鹏、黄海荣、黄杏、唐仕云、李翔、陈荣发 Yu-chi Deng, Lun-wang Wang, Wu Xian, Jian-ming Wu, Xiu-peng Song, Hai-rong Huang, Xing Huang, Shi-yun Tang, Xiang Li, Rong-fa Chen
43	一种糖度检测装置盛装袋 A packing bag for sugar testing device	2019-04-29	201920599391.1	实用新型 Utility model	张保青、杨翠芳、周珊、周忠凤、段维兴、张革民、黄玉新、王艳萍 Bao-qing Zhang, Cui-fang Yang, Shan Zhou, Zhong-feng Zhou, Wei-xing Duan, Ge-min Zhang, Yu-xin Huang, Yan-ping Wang

2.2.2 授权专利 Authorized Patents

2019年获授权专利（表7）。

Patents authorized in 2019 (Table 7).

表7 2019年获授权专利

Table 7 Patents authorized in 2019

序号 No.	专利名称 Patent Title	申请时间 Application Time	授权时间 Authorized Date	专利号 Patent No.	专利类型 Patent Type	发明人 Inventor
1	一种甘蔗实生苗除草剂及其使用方法 A sugarcane seedling herbicide and its use method	2016-11-02	2019-03-22	ZL201610942140.X	发明专利 Invention	王泽平，张保青，周珊，杨翠芳，高铁静，罗霆，韦金菊，段维兴 Ze-ping Wang, Bao-qing Zhang, Ge-min Zhang, Shan Zhou, Cui-fang Yang, Yi-jing Gao, Ting Luo, Wei-xing Duan
2	一种甘蔗叶片中柠檬酸含量的HPLC检测方法的应用 Application of the HPLC method for determining citric acid content in sugarcane leave	2016-10-14	2019-06-07	ZL201610896893.1	发明专利 Invention	王泽平，李长宁，刘璐，张保青，杨翠芳，高铁静，蒋洪涛，罗霆，梁强，林善海，段维兴，张保青，周珊，韦金菊，张保青，常宁丽，卢柳，张保青 Ze-ping Wang, Chang-ning Li, Lu Liu, Bao-qing Zhang, Cui-fang Yang, Yi-jing Gao, Hong-tao Jiang, Ting Luo, Qiang Liang, Shan-hai Lin, Wei-xing Duan, Ge-min Zhang, Shan Zhou, Jin-ju Wei
3	一种甘蔗叶片中苹果酸含量的HPLC检测方法 A HPLC method for determining apple acid content in sugarcane leaves	2016-10-14	2019-06-07	ZL201610897040.X	发明专利 Invention	王泽平，梁强，周珊，罗霆，韦金菊，刘璐，张保青，高铁静，杨翠芳，林波，蒋洪涛，段维兴 Ze-ping Wang, Qiang Liang, Shan-hai Lin, Shan Zhou, Ting Luo, Jin-ju Wei, Lu Liu, Ge-min Zhang, Yi-jing Gao, Cui-fang Yang, Bao-qing Zhang, Bo Lin, Hong-tao Jiang, Wei-xing Duan
4	一种甘蔗叶片中茉莉酸含量的HPLC检测方法 A HPLC method for determining jasmine acid content in sugarcane leaves	2016-10-14	2019-04-26	CN201610896902.7	发明专利 Invention	王泽平，林善海，蒋洪涛，林波，周珊，何为中，李毅杰，罗霆，杨翠芳，韦金菊，高铁静，张保青，刘维兴 Ze-ping Wang, Shan-hai Lin, Hong-tao Jiang, Bo Lin, Shan Zhou, Wei-zhong He, Yi-jie Li, Ting Luo, Cui-fang Yang, Jin-ju Wei, Yi-jing Gao, Ge-min Zhang, Lu Liu, Wei-xing Duan

(续)

序号 No.	专利名称 Patent Title	申请时间 Application Time	授权时间 Authorized Date	专利号 Patent No.	专利类型 Patent Type	发明人 Inventor
5	一种甘蔗叶片中赤霉素含量的HPLC检测方法 A HPLC method for determining erythromycin content in sugarcane leaves	2016-10-14	2019-04-26	CN201610897056.0	发明专利 Invention	王泽平，刘璐，李毅杰，段维兴，梁强，张革民，罗霆，韦金菊，高铁静，林波，周珊，杨翠芳，林善海，张保青 Ze-ping Wang, Lu Liu, Yi-jie Li, Wei-xing Duan, Qiang Liang, Ge-min Zhang, Ting Luo, Jin-ju Wei, Yi-jing Gao, Bo Lin, Shan Zhou, Cui-fang Yang, Shan-hai Lin, Bao-qing Zhang
6	一种甘蔗叶片中水杨酸含量的HPLC检测方法 A HPLC method for determining salicylic acid content in sugarcane leaves	2016-10-14	2019-07-23	CN201610896901.2	发明专利 Invention	王泽平，李毅杰，刘璐，蒋洪涛，李长宁，张保青，高铁静，罗霆，张革民，韦金菊，段维兴，周珊，杨翠芳，林善海 Ze-ping Wang, Yi-jie Li, Lu Liu, Hong-tao Jiang, Chang-ning Li, Bao-qing Zhang, Yi-jing Gao, Ting Luo, Ge-min Zhang, Jin-ju Wei, Wei-xing Duan, Shan Zhou, Cui-fang Yang, Shan-hai Lin
7	根癌农杆菌介导的荸荠秆枯病菌的遗传转化方法 A genetic transformation method for Cylindrosporium eleocharidis mediated by Agrobacterium tumefaciens	2016-07-08	2019-06-21	CN 105969799 B	发明专利 Invention	颜梅新，江文，欧昆鹏，桂杰，董伟清，高美萍，何芳练，张尚文 Mei-xin Yan, Wen Jiang, Kun-peng Ou, Jie gui, Wei-qing Dong, Mei-ping Gao, Fang-lian He, Shang-wen Zhang
8	一种纯化细菌污染丝状真菌的简易方法 An easy way to purify filamentous fungi contaminated by bacteria	2015-12-25	2019-05-28	CN 105462856 B	发明专利 Invention	林善海，黄诚华，周主贵，商显坤，潘雪红，魏吉利，尚学红，王泽平，李毅杰 Shan-hai Lin, Cheng-hua Huang, Zhu-gui Zhou, Xian-kun Shang, Xue-hong Pan, Ji-li Wei, Ze-ping Wang, Yi-jie Li
9	一种从甘蔗叶片上分离赤腐病菌的简易方法 An easy method for separating Colletotrichum falcatum from sugarcane leaves	2015-12-28	2019-05-28	CN 105441335 B	发明专利 Invention	林善海，黄诚华，王泽平，周主贵，魏吉利，潘雪红，商显坤 Shan-hai Lin, Cheng-hua Huang, Yi-jie Li, Ze-ping Wang, Zhu-gui Zhou, Ji-li Wei, Xue-hong Pan, Xian-Kun Shang
10	一种定量刺伤并接种菌液的接种器 A quantitative stab wound inoculator for bacteria solution	2018-10-23	2019-07-09	ZL201821721976.8	实用新型 Utility model	韦金菊，宋修鹏，张小秋，魏春燕，覃振强，颜梅新，刘昔辉，张荣华，桂意云，李意云，周会 Jin-ju Wei, Xiu-peng Song, Xiao-qiu Zhang, Chun-yan Wei, Zhen-qiang Qin, Mei-xin Yan, Xi-hui Liu, Rong-hua Zhang, Yi-yun Gui, Hai-bi Li, Hui Zhou

(续)

序号 No.	专利名称 Patent Title	申请时间 Application Time	授权时间 Authorized Date	专利号 Patent No.	专利类型 Patent Type	发明人 Inventor
11	一种甘蔗黑穗病孢子粉收集器 A collector for sugarcane smut spore	2018-10-23	2019-07-09	ZL201821723654.7	实用新型 Utility model	韦金菊、宋修鹏、张小秋、魏春燕、覃振强、颜梅新、刘昔辉、张荣华、桂意云、李海碧、李杨瑞 Jin-ju Wei, Xiu-peng Song, Xiao-qiu Zhang, Chun-yan Wei, Zhen-qiang Qin, Mei-xin Yan, Xi-hui Liu, Rong-hua Zhang, Yi-yun Gui, Hai-bi Li, Yang-rui Li
12	一种收集锈病孢子粉的收集器 A collector for rust spores	2018-10-23	2019-07-09	ZL201821721758.4	实用新型 Utility model	韦金菊、周会、宋修鹏、张荣华、魏春燕、桂意云、李海碧、颜梅新、刘昔辉、张小秋、覃振强、李杨瑞 Jin-ju Wei, Hui Zhou, Xiu-peng Song, Xiao-qiu Zhang, Chun-yan Wei, Zhen-qiang Qin, Mei-xin Yan, Xi-hui Liu, Rong-hua Zhang, Yi-yun Gui, Hai-bi Li, Yang-rui Li
13	一种接种黑穗病孢子菌液的针 A needle for inoculation with smut spore liquid	2018-10-23	2019-07-16	ZL201821720733.2	实用新型 Utility model	韦金菊、宋修鹏、张小秋、魏春燕、桂意云、李海碧、刘昔辉、张荣华、颜梅新、覃振强、李杨瑞 Jin-ju Wei, Xiu-peng Song, Xiao-qiu Zhang, Chun-yan Wei, Zhen-qiang Qin, Mei-xin Yan, Xi-hui Liu, Rong-hua Zhang, Yi-yun Gui, Hai-bi Li, Hui Zhou, Yang-rui Li
14	一种螺旋挤压式甘蔗取汁装置 A spiral squeeze cane juicer	2018-10-23	2019-07-16	ZL201821720470.5	实用新型 Utility model	韦金菊、李海碧、张小秋、宋修鹏、周会、王维赞、覃振强 Jin-ju Wei, Hai-bi Li, Xiao-qiu Zhang, Rong-hua Zhang, Yi-yun Gui, Yang-rui Li, Wei-zan Wang, Zhen-qiang Qin, Xiu-peng Song, Hui Zhou
15	一种直立作物测高装置 A device for measuring upright crop height	2018-11-01	2019-06-25	ZL201821792599.7	实用新型 Utility model	韦春燕、张荣华、刘昔辉、张小秋、桂意云、李海碧、黄东亮 Jin-ju Wei, Rong-hua Zhang, Xi-hui Liu, Xiao-qiu Zhang, Yi-yun Gui, Hai-bi Li, Chun-yan Wei, Hui Zhou, Yang-rui Li, Dong-liang Huang
16	一种自动排种式甘蔗种植机 A sugarcane growing machine with automatic arranging cane seeds capacity	2018-12-28	2019-11-01	201822232806.X	实用新型 Utility model	庞天、覃建才、王维赞、李毅杰、张荣华、吴凯朝、罗亚伟、刘晓燕、覃文宪、邓智年、黄成丰 Tian Pang, Jian-cai Qin, Wei-zan Wang, Yi-jie Li, Rong-hua Zhang, Kai-chao Wu, Ya-wei Luo, Xiao-yan Liu, Wen-xian Qin, Zhi-nian Deng, Cheng-feng Huang

(续)

序号 No.	专利名称 Patent Title	申请时间 Application Time	授权时间 Authorized Date	专利号 Patent No.	专利类型 Patent Type	发明人 Inventor
17	一种液态肥同步施肥机 A synchronous fertilizing machine for liquid fertilizer	2018-12-28	2019-11-01	201822232809.9	实用新型 Utility model	邓智年，覃建才，李毅杰，张荣华，庞天，王维赞，吴凯朝，罗亚伟，刘晓燕，覃文凭，黄成丰 Zhi-nian Deng, Jian-cai Qin, Yi-jie Li, Rong-hua Zhang, Tian Pang, Wei-zan Wang, Kai-chao Wu, Ya-wei Luo, Xiao-yan Liu, Wen-xian Qin, Cheng-feng Huang
18	一种平地机 A machine for field leveling	2018-12-28	2019-11-01	201822232871.2	实用新型 Utility model	庞天，覃建才，王维赞，张荣华，邓智年，吴凯朝，李毅杰，罗亚伟，刘晓燕，覃文凭，黄成丰 Tian Pang, Jian-cai Qin, Wei-zan Wang, Rong-hua Zhang, Zhi-nian Deng, Kai-chao Wu, Yi-jie Li, Ya-wei Luo, Xiao-yan Liu, Wen-xian Qin, Cheng-feng Huang
19	一种可调式甘蔗培土机 An adjustable sugarcane earthing up machine	2018-12-28	2019-10-09	201822231288.X	实用新型 Utility model	王维赞，覃建才，庞天，邓智年，吴凯朝，周慧文，梁阗，覃文凭 Wei-zan Wang, Jian-cai Qin, Tian Pang, Zhi-nian Deng, Rong-hua Zhang, Yi-jie Li, Kai-chao Wu, Hui-wen Zhou, Tian Liang, Wen-xian Qin
20	一种深松器 A deep plowing and scarification machine	2018-12-23	2019-10-08	201822231336.5	实用新型 Utility model	王维赞，刘晓燕，庞天，李毅杰，吴凯朝，罗亚伟，覃文凭，邓智年 Wei-zan Wang, Jian-cai Qin, Tian Pang, Yi-jie Li, Rong-hua Zhang, Kai-chao Wu, Ya-wei Luo, Xiao-yan Liu, Wen-xian Qin, Zhi-nian Deng
21	一种高效可控的精准电动甘蔗种茎切种机 An efficient, controlable and precise electric sugarcane stem cutting machine	2018-05-04	2019-01-01	ZL200182065934Z0	实用新型 Utility model	张荣华，庞天，邓智年，李毅杰，吴凯朝，覃文凭，梁阗，周慧文 Rong-hua Zhang, Tian Pang, Zhi-nian Deng, Yi-jie Li, Kai-chao Wu, Wei-zan Wang, Hui-wen Zhou, Tian Liang, Wen-xian Qin
22	一种手动精准甘蔗种茎切种机 A manual precision sugarcane stem cutting machine	2018-05-04	2019-01-01	ZL201820657794.2	实用新型 Utility model	庞天，王维赞，李毅杰，邓智年，吴凯朝，周慧文，张荣华，覃文凭 Tian Pang, Wei-zan Wang, Yi-jie Li, Zhi-nian Deng, Kai-chao Wu, Hui-wen Zhou, Rong-hua Zhang, Wen-xian Qin
23	一种用于挤压式甘蔗切种的特型刀具 A special extrusive tool for sugarcane seed cutting	2018-07-19	2019-06-25	ZL201821147768.1	实用新型 Utility model	庞天，刘晓燕，邓智年，王维赞，罗亚伟，覃文凭 Tian Pang, Yi-jie Li, Rong-hua Zhang, Kai-chao Wu, Zhi-nian Deng, Wei-zan Wang, Ya-wei Luo, Xiao-yan Liu, Wen-xian Qin

(续)

序号 No.	专利名称 Patent Title	申请时间 Application Time	授权时间 Authorized Date	专利号 Patent No.	专利类型 Patent Type	发明人 Inventor
24	一种高效可控的精准气动甘蔗种茎切种机 A highly efficient and controlable precision pneumatic sugarcane stem cutting machine	2018-05-04	2019-01-01	ZL201820687753.3	实用新型 Utility model	王维赞,邓智年,李毅杰,张荣华,庞天,吴凯朝,罗亚伟,刘晓燕,覃文宪 Wei-zan Wang, Zhi-nian Deng, Yi-jie Li, Rong-hua Zhang, Tian Pang, Kai-chao Wu, Ya-wei Luo, Xiao-yan Liu, Wen-xian Qin
25	一种滑机式甘蔗切种装置 A rail-type sugarcane cutting device	2018-07-19	2019-07-02	ZL201821146737.4	实用新型 Utility model	张荣华,庞天,王维赞,李毅杰,邓智年,吴凯朝,周慧文,覃文宪 Rong-hua Zhang, Tian Pang, Wei-zan Wang, Yi-jie Li, Zhi-nian Deng, Kai-chao Wu, Hui-wen Zhou, Wen-xian Qin
26	一种观察甘蔗开花及采集数据的装置 A device for observing sugarcane flowering and collecting data	2018-11-29	2019-06-25	ZL201821986307.3	实用新型 Utility model	张荣华,庞天,邓智年,王维赞,黄成丰 Rong-hua Zhang, Tian Pang, Zhi-nian Deng, Wei-zan Wang, Cheng-feng Huang
27	一种甘蔗防倒伏装置 A device for protecing sugarcane lodging	2018-08-31	2019-06-18	ZL201821421700.8	实用新型 Utility model	罗亚伟,覃振强,李德伟,宋修鹏,韦金菊,张小平 Ya-wei Luo, Zheng-qiang Qin, De-wei Li, Xiu-peng Song, Jin-ju Wei, Chun-yan Wei, Xiao-ping Zhang
28	一种多功能喷淋车 A multi-purpose sprinkler	2018-08-31	2019-06-18	ZL201821422690.X	实用新型 Utility model	罗亚伟,覃振强,李德伟,张小平,韦金菊 Ya-wei Luo, Zhen-qiang Qin, De-wei Li, Xiao-ping Zhang, Xiu-peng Song, Chun-yan Wei, Jin-ju Wei
29	一种甘蔗除草平地轮 A wheel for sugarcane weeding	2018-08-31	2019-06-18	ZL201821423591.3	实用新型 Utility model	罗亚伟,覃振强,李德伟,张小平,韦金菊,宋修鹏,魏春燕 Ya-wei Luo, Zhen-qiang Qin, De-wei Li, Xiao-ping Zhang, Jin-ju Wei, Xiu-peng Song, Chun-yan Wei
30	一种可拆装喷淋系统 A removable spray system	2018-08-31	2019-06-14	ZL201821423594.7	实用新型 Utility model	罗亚伟,覃振强,李德伟,宋修鹏,韦金菊,梁阗,魏春燕,张小平 Ya-wei Luo, Zhen-qiang Qin, De-wei Li, Xiu-peng Song, Jin-ju Wei, Tian Liang, Chun-yan Wei, Xiao-ping Zhang

序号 No.	专利名称 Patent Title	申请时间 Application Time	授权时间 Authorized Date	专利号 Patent No.	专利类型 Patent Type	发明人 Inventor
31	一种作物抗旱运水车 A water tanker for protecting crop against drought damage	2018-08-31	2019-06-14	ZL201821422642.0	实用新型 Utility model	罗亚伟、覃振强、李德伟、梁阗、张小平、魏春燕、宋修鹏、韦金菊 Ya-wei Luo, Zhen-qiang Qin, De-wei Li, Tian Liang, Xiao-ping Zhang, Chun-yan Wei, Xiu-peng Song, Jin-ju Wei
32	一种甘蔗抗旱漫灌系统 A irrigation system for protecting sugarcane against drought damage	2018-08-31	2019-07-09	ZL201821421683.8	实用新型 Utility model	罗亚伟、覃振强、李德伟、魏春燕、韦金菊、宋修鹏、梁阗、张小平 Ya-wei Luo, Zhen-qiang Qin, De-wei Li, Chun-yan Wei, Jin-ju Wei, Xiu-peng Song, Tian Liang, Xiao-ping Zhang
33	一种农药雾滴沉积检测卡固定装置 A device for fixing pesticide mist drop detection card	2018-12-13	2019-07-03	201822086961.5	实用新型 Utility model	宋修鹏、张小秋、梁永检、李杨瑞、魏春燕、颜梅新、覃振强、宋奇琦、韦金菊、李立涛、秦振强、颜春燕、李德伟 Xiu-peng Song, Xiao-qiu Zhang, Yong-jian Liang, Yang-rui Li, Li-tao Yang, Zhen-qiang Qin, Qi-qi Song, Jin-ju Wei, Mei-xin Yan, Chun-yan Wei, De-wei Li
34	一种甘蔗实生苗培育装置 A device for sugarcane seedling	2019-03-11	2019-10-10	201920302794.5	实用新型 Utility model	邓宇驰、王伦旺、唐仕云、周忠凤 Yu-chi Deng, Lun-wang Wang, Shi-yun Tang, Zhong-feng Zhou
35	一种甘蔗转运车 A vehicle for sugarcane transport	2019-03-21	2019-11-04	201920362026.9	实用新型 Utility model	邓宇驰、雷敬超、黄海荣、杨荣仲、高丽花 Yu-chi Deng, Jing-chao Lei, Hai-rong Huang, Rong-zhong Yang, Li-hua Gao
36	一种甘蔗茎装置 A device for stem sugarcane	2019-03-21	2019-10-31	201920362647.7	实用新型 Utility model	邓宇驰、贤武、王伦旺、雷敬超、高丽花 Yu-chi Deng, Wu Xian, Lun-wang Wang, Jing-chao Lei, Li-hua Gao
37	一种甘蔗实生苗转运装置 A device for transporting sugarcane seedling	2019-03-21	2019-10-18	201920362648.1	实用新型 Utility model	邓宇驰、王伦旺、唐仕云、杨荣仲 Yu-chi Deng, Lun-wang Wang, Shi-yun Tang, Rong-zhong Yang
38	一种甘蔗取汁器 A device for sampling sugarcane juice	2019-01-25	2019-09-26	201920132219.5	实用新型 Utility model	王伦旺、贤武、经艳、黄海荣 Lun-wang Wang, Wu Xian, Yan Jing, Hai-rong Huang
39	一种甘蔗电锯刀 A electric saw knife for sugarcane	2019-03-27	2019-11-06	201920400525.2	实用新型 Utility model	邓宇驰、黄海荣、王伦旺、贤武、谭芳、杨荣仲、雷敬超 Yu-chi Deng, Hai-rong Huang, Lun-wang Wang, Wu Xian, Fang Tan, Rong-zhong Yang, Jing-chao Lei

(续)

序号 No.	专利名称 Patent Title	申请时间 Application Time	授权时间 Authorized Date	专利号 Patent No.	专利类型 Patent Type	发明人 Inventor
40	一种甘蔗剥叶工具 A tool for peeling sugarcane leaves	2019-01-25	2019-09-04	201920131240.3	实用新型 Utility model	贤武，王伦旺，经艳，谭芳 Wu Xian, Lun-wang Wang, Yan Jing, Fang Tang
41	一种气动挤压式自动高效精准可控甘蔗切种机 A pneumatic extrusive, automatic, high-efficiency, precise and controllable sugarcane seed cutting machine	2018-07-19	2019-06-25	ZL201821147772.8	实用新型 Utility model	邓智年，张荣华，庞天，李毅杰，吴凯朝，王维赞，周慧文，梁阗，覃文芫 Zhi-nian Deng, Rong-hua Zhang, Tian Pang, Yi-jie Li, Kai-chao Wu, Wei-zan Wang, Hui-wen Zhou, Tian Liang, Wen-xian Qin
42	一种诱捕盆自动补水装置 A trap basin with automatic water-adding device	2019-04-28	2019-11-14	201920531096.2	实用新型 Utility model	黄诚华，商显坤，魏吉利，潘雪红 Cheng-hua Huang, Xian-kun Shang, Ji-li Wei, Xue-hong Pan
44	一种起垄种植式甘蔗抗倒伏支架 A bracket for protecting ridge-growing sugarcane against lodging	2019-04-18	2019-08-30	ZL201821836163.3	实用新型 Utility model	刘俊仙，熊发前，李松，雷敬超，丘立杭，刘菁，刘丽敏，段维兴，刘红坚，卢曼曼，何毅波，张伟珍 Jun-xian Liu, Fa-qian Xiong, Song Li, Jing-chao Lei, Li-hang Qiu, Jing Liu, Li-min Liu, Wei-xing Duan, Hong-jian Liu, Man-man Lu, Yi-bo He, Wei-zhen Zhang
45	一种用于组织培养瓶的超声波清洗装置 An ultrasonic cleaning device for tissue culture bottles	2018-11-08	2019-03-19	ZL201822042904.0	实用新型 Utility model	刘俊仙，李松，熊发前，刘菁，刘丽敏，吴建明，丘立杭，何毅波，张伟珍 Jun-xian Liu, Song Li, Fa-qian Xiong, Jing Liu, Li-min Liu, Hong-jian Liu, Man-man Lu, Wei-zhong He, Jian-ming Wu, Li-hang Qiu, Yi-bo He, Wei-zhen Zhang
46	一种植物组织培养瓶的自动清洗烘干一体装置 A devise with automatic cleaning and drying capacity for plant tissue culture bottles	2018-07-03	2019-07-23	ZL201821043524.9	实用新型 Utility model	刘俊仙，何为中，吴建明，刘菁，刘红坚，熊发前，卢曼曼，张伟珍，刘丽敏，丘立杭 Jun-xian Liu, Li-min Liu, Li-hang Qiu, Jing Liu, Hong-jian Liu, Fa-qian Xiong, Wei-zhong He, Jian-ming Wu, Man-man Lu, Wei-zhen Zhang, Yi-bo He
47	一种适于室外甘蔗抗旱试验的挡雨罩 A rain shield suitable for outdoor sugarcane drought-resistant test	2018-10-23	2019-10-01	CN 209457439.U	实用新型 Utility model	桂意云，张荣华，刘昔辉，刘晓燕，林丽，周会，李杨瑞，廖芬，刘忠良，陈忠良，李林，李杨瑞 Yi-yun Gui, Rong-hua Zhang, Xi-hui Liu, Jian-ming Wu, Zhong-liang Chen, Fen Liao, Xiao-yan Liu, Li Lin, Hui Zhou, Yang-rui Li

2.3 软件著作权 Software Copyright

2019年登记软件著作权（表8）。

Software copyrights registered in 2019 (Table 8).

表8 2019年登记软件著作权

Table 8 Software copyrights registered in 2019

序号 No.	软件名称 Title	获得时间 Registered Date	著作权编号 Copyright No.	著作权人 Owners
1	固氮基因 NIFH 的预测及分析软件 Software for prediction and analysis of nitrogen-fixing gene *nifH*	2019-08-02	2019SR1094162	李长宁，农倩，林丽，莫璋红，谢金兰，罗霆 Chang-ning Li, Qian Nong, Li Lin, Zhang-hong Mo, Jin-lan Xie, Ting Luo
2	甘蔗 EST 序列的 SNP 标记开发及分析软件 Software for development and analysis of SNP marker from sugarcane EST sequence	2019-07-21	2019SR0947142	李长宁，农倩，林丽，罗霆，谢金兰，莫璋红 Chang-ning Li, Qian Nong, Li Lin, Ting Luo, Jin-lan Xie, Zhang-hong Mo
3	甘蔗 EST 序列的 SSR 标记开发及分析软件 Software for development and analysis of SSR marker from sugarcane EST sequence	2019-06-12	2019SR0931220	李长宁，农倩，林丽，罗霆，谢金兰，莫璋红 Chang-ning Li, Qian Nong, Li Lin, Ting Luo, Jin-lan Xie, Zhang-hong Mo

2.4 标准 Standard

2019年发布的标准（表9）。

Standards released in 2019 (Table 9).

表9 2019年发布的标准

Table 9　Standards released in 2019

序号 No.	名称 Title	类型 Type	标准号 Standard No.	发布日期 Released Date	起草人 Authors
1	甘蔗DNA指纹图谱采集技术规程 Technical regulations for the collection of sugarcane DNA fingerprint	广西地方标准 Guangxi Local Standard	DB45/T 2026—2019	2019-12-30	高铁静，周会，杨来仲，张来华，段维兴，张保青，杨翠芳，周珊，王泽平，张革民 Yi-jing Gao, Hui Zhou, Rong-zhong Yang, Fa-qian Xiong, Wei-xing Duan, Bao-qing Zhang, Cui-fang Yang, Shan Zhou, Ze-ping Wang, Ge-min Zhang
2	甘蔗粉蚧为害调查技术规范 Technical regulations for sugarcane damage caused by mealybugs	广西地方标准 Guangxi Local Standard	DB45/T 2027—2019	2019-12-30	覃振强，李德伟，王全永，黄珽，宋修鹏，魏春燕，韦金菊，张燕杏，温丽玲，唐艳琼 Zhen-qiang Qin, De-wei Li, Quan-yong Wang, Xiu-peng Song, Chun-yan Wei, Jin-ju Wei, Ya-wei Luo, Yan-xing Zhang, Li-ling Wen, Kun Huang, Yan-qiong Tang
3	甘蔗组织培养污染控制操作规程 Technical regulations for aseptic operation procedure of sugarcane tissue culture	广西地方标准 Guangxi Local Standard	DB45/T 2028—2019	2019-12-30	李松，刘红坚，卢曼曼，刘俊仙，刘丽敏，张荣华，余坤兴，何毅波，刘欣，张伟珍 Song Li, Hong-jian Liu, Man-man Lu, Jun-xian Liu, Li-min Liu, Rong-hua Zhang, Kun-xing Yu, Yi-bo He, Xin Liu, Wei-zhen Zhang
4	甘蔗白条病菌PCR检测技术规程 Detecting technique regulations for pathogen of *Xanthomonas albilineans* (Ashby) Downson based on PCR	广西地方标准 Guangxi Local Standard	DB45/T 2029—2019	2019-12-30	韦金菊，刘昔辉，覃振强，张小秋，张来华，桂意云，魏春燕，宋修鹏，李杨瑞，周珊，黄东亮，颜梅新，刘璐 Jin-ju Wei, Xi-hui Liu, Zhen-qiang Qin, Xiao-qiu Zhang, Rong-hua Zhang, Yi-yun Gui, Chun-yan Wei, Xiu-peng Song, Yang-rui Li, Shan Zhou, Dong-liang Huang, Mei-xin Yan, Lu Liu
5	甘蔗粉蚧防治技术规程 Control technology procedure for sugarcane mealybugs	广西地方标准 Guangxi Local Standard	DB45/T 2030—2019	2019-12-30	覃振强，李德伟，罗亚伟，宋修鹏，魏春燕，韦金菊，张燕杏，温丽玲，彭应知，黄珽 Zhen-qiang Qin, De-wei Li, Ya-wei Luo, Quan-yong Wang, Xiu-peng Song, Chun-yan Wei, Jin-ju Wei, Yan-xing Zhang, Li-ling Wen, Ying-zhi Peng, Kun Huang

(续)

序号 No.	名称 Title	类型 Type	标准号 Standard No.	发布日期 Released Date	起草人 Authors
6	甘蔗白叶病植原体巢式PCR检测技术规程 Detecting technique regulations for phytoplasma of sugarcane wild leaf based on nested PCR	广西地方标准 Guangxi Local Standard	DB45/T 2031—2019	2019-12-30	韦金菊，宋修鹏，刘昔辉，覃振强，李杨瑞，张小秋，李杨瑞，张荣华，桂意云，魏春燕，刘丽敏，吴建明，颜梅新，刘璐 Jin-ju Wei, Xiu-peng Song, Xi-hui Liu, Zhen-qiang Qin, Xiao-qiu Zhang, Yang-rui Li, Rong-hua Zhang, Yi-yun Gui, Chun-yan Wei, Li-min Liu, Jian-ming Wu, Mei-xin Yan, Lu Liu
7	糖料甘蔗良种逐级繁育技术规程 Technical specification for production of three grade cultivar-multiplication of sugarcane	广西地方标准 Guangxi Local Standard	DB45/T 2032—2019	2019-12-30	李松，庞天，邓智年，王维赞，刘俊仙，刘红坚，刘丽敏，卢曼曼，何毅波，张荣华，覃振强，林善海 Song Li, Tian Pang, Zhi-nian Deng, Wei-zan Wang, Jun-xian Liu, Hong-jian Liu, Li-min Liu, Man-man Lu, Yi-bo He, Rong-hua Zhang, Zhen-qiang Qin, Shan-hai Lin
8	甘蔗试管苗光合自养生根技术规程 Specification for photoautotrophic rooting of sugarcane microshoots	广西地方标准 Guangxi Local Standard	DB45/T 2033—2019	2019-12-30	何为中，刘丽敏，梁阗，刘红坚，翁梦苓，谢金兰 Wei-zhong He, Li-min Liu, Tian Liang, Hong-jian Liu, Meng-ling Weng, Jin-lan Xie
9	甘蔗茎尖脱毒组培苗繁育技术规程 Technical specification for sugarcane virus-free tissue culture seedlings from shoot tip rapid propagation	广西地方标准 Guangxi Local Standard	DB45/T 2034—2019	2019-12-30	刘丽敏，刘红坚，李松，何为中，刘俊仙，翁梦苓，卢曼曼，何毅波，张伟珍 Li-min Liu, Hong-jian Liu, Song Li, Wei-zhong He, Jun-xian Liu, Meng-ling Weng, Man-man Lu, Yi-bo He, Wei-zhen Zhang

3 发表论著
PUBLICATIONS

3.1 发表期刊论文 Journal Papers

1. Shu-quan Sun, Yi-zhen Deng, En-ping Cai, Mei-xin Yan, Ling-yu Li, Bao-shan Chen, Chang-qing Chang*, Zi-de Jiang*. The farnesyltransferase β-subunit ram1 regulates *Sporisorium scitamineum* mating, pathogenicity and cell wall integrity. Frontiers in microbiology, 2019, 10:976.

2. Hai-feng Yan, Ming-zhi Li, Yu-ping Xiong, Jian-ming Wu*, Da Silva Jaime A. Teixeira*, Guo-hua Ma*. Genome-Wide Characterization, Expression profile analysis of *WRKY* family genes in *Santalum album* and functional identification of their role in abiotic stress. International Journal of Molecular Sciences, 2019, 20(22):5676.

3. Li-hang Qiu, Rong-fa Chen, Ye-geng Fan, Xing Huang, Han-min Luo, Fa-qian Xiong, Jun-xian Liu, Rong-hua Zhang, Jing-chao Lei, Hui-wen Zhou, Jian-ming Wu*, Yang-rui Li*. Integrated mRNA and small RNA sequencing reveals microRNA regulatory network associated with internode elongation in sugarcane (Saccharum officinarum L.). BMC Genomics, 2019, 20(1):1-14.

4. Yue Zhong, Mei-xin Yan, Jin-yan Jiang, Zhi-han Zhang, Jun-jun Huang, Lian-hui Zhang, Yin-yue Deng*, Xiao-fan Zhou*, Fei He*. Mycophenolic acid as a promising fungal dimorphism inhibitor to control sugarcane disease caused by *Sporisorium scitamineum*. Journal of agricultural and food chemistry, 2019, 67(1):112-119.

5. Zhong-liang Chen, Cui-xian Qin, Miao Wang, Fen Liao, Qing Liao, Xi-hui Liu, Yang-rui Li, Prakash Lakshmanan, Ming-hua Long*, Dong-liang Huang*. Ethylene-mediated improvement in sucrose accumulation in ripening sugarcane involves increased sink strength. BMC Plant Biology, 2019, 19(1):1-17.

6. Kai Zhu, Min Shao, Dan Zhou, Yong-xiu Xing, Li-tao Yang*, Yang-rui Li*. Functional analysis of *Leifsonia xyli* subsp. *xyli* membrane protein gene *Lxx18460* (anti-sigma K). BMC Microbiology, 2019, 19(1):2.

7. Qian-qian You, Juan Peng, Yi-peng Xie, Xin Li, Zhi-guang Tang, Bin Feng, Li-jun Wang*, Meng-ling Weng*, Gan-lin Chen*. Optimization of protective agents and freeze-drying conditions for *Lactobacillus casei* LT-L614. Journal of Biobased Materials and Bioenergy, 2019, 13(1):123-128.

8. Juan Peng, Qian-qian You, Bin Feng, Zhi-neng Chen, Yong-hong Li, Zhi-guang Tang, Xin Li, Li-jun Wang*, Jian Chen, Meng-ling Weng*. Cultivation of Lactobacillus casei LT-L614 in High density with sugarcane molasses: Optimization of processing. Journal of Biobased Materials and Bioenergy, 2019, DOI:10.1166/jbmb.2019.1920.

9. Manoj Kumar Solanki, Fei-yong Wang, Zhen Wang, Chang-ning Li, Tao-ju Lan, Rajesh Kumar Singh, Pratiksha Singh, Li-tao Yang, Yang-rui Li*. Rhizospheric and endospheric diazotrophs mediated soil fertility intensification in sugarcane-legume intercropping systems. Journal of Soils & Sediments: Protection, Risk Assessment, & Remediation, 2019, 19(4):1911-1927.

10. Ze-ping Wang, Yi-jie Li, Chang-ning Li, Xiu-peng Song, Jing-chao Lei, Yi-jing Gao, Qiang Liang*. Comparative transcriptome profiling of resistant and susceptible sugarcane genotypes in response to the airborne pathogen Fusarium verticillioides. Molecular Biology Reports, 2019, 46(4):3777-3789.

11. Zhen-qiang Qin*, Jian-hui Wu, Bao-li Qiu, Shaukat Ali, Andrew G S Cuthbertson. The impact of *Cryptolaemus montrouzieri* Mulsant (Coleoptera: Coccinellidae) on control of *Dysmicoccus neobrevipes* Beardsley (Hemiptera: Pseudococcidae). Insects, 2019, 10(5):131.

12. K.K. Verma, Xi-hui Liu, Kai-chao Wu, R.K. Singh, Qi-qi Song, M.K. Malviya, Xiu-peng Song, P. Singh, C.L. Verma, Yang-rui Li*. The impact of silicon on photosynthetic and biochemical responses of sugarcane under different soil moisture levels. Silicon,2019,doi:10.1007/s12633-019-00228-z.

13. Li-hang Qiu, Rong-fa Chen, Han-min Luo, Ye-geng Fan, Xing Huang, Jun-xian Liu, Fa-qian Xiong, Hui-wen Zhou, Chong-kun Gan, Jian-ming Wu*, Yang-rui Li*. Effects of exogenous GA_3 and DPC treatments on levels of endogenous hormone and expression of key gibberellin biosynthesis pathway genes during stem elongation in sugarcane. Sugar Tech, 2019, 21(6):936-948.

14. Bao-qing Zhang, Min Shao, Yong-jian Liang, Xing Huang, Xiu-peng Song, Hui Chen, Li-tao Yang*, Yang-rui Li*. Molecular cloning and expression analysis of *ScTUA* gene in sugarcane. Sugar Tech, 2019, 21(4):578-585.

15. Xiao-qiu Zhang, Yong-jian Liang, Zhen-qiang Qin, De-wei Li, Chun-yan Wei, Jin-ju Wei, Yang-rui Li*, Xiu-peng Song*. Application of multi-rotor unmanned aerial vehicle application in management of stem borer (Lepidoptera) in sugarcane. Sugar Tech, 2019, 21(5):847-852.

16. Yong-jian Liang, Xiao-qiu Zhang, Liu Yang, Xi-hui Liu, Li-tao Yang*, Yang-rui Li*. Impact of seed coating agents on single-bud seedcane germination and plant growth in commercial sugarcane cultivation. Sugar Tech, 2019, 21 (3):383-387.

17. M.K. Solanki, Fei-yong Wang, Chang-ning Li, Zhen Wang, Tao-ju Lan, Singh R.K., P. Singh, Li-tao Yang, Yang-rui Li*. Impact of sugarcane–legume intercropping on diazotrophic microbiome. Sugar Tech, 2019, doi:10.1007/ s12355-019-00755-4.

18. Jun-qi Niu, Jing-li Huang, Thithu Phan, Yong-bao Pan, Li-tao Yang, Yang-rui Li*. Molecular cloning and expressional analysis of five sucrose transporter (*SUT*) genes in sugarcane. Sugar

Tech, 2019, 21(1):47-54.

19. Fan-wei Wang, Jian-bin Yu, Dong Xiao, Yang-rui Li*, Long-fei He*, ;Ai-qin Wang*. Molecular cloning and expression analysis of ethylene-insensitive3-like 1 (*ScEIL1*) gene in sugarcane. Sugar Tech, 2019, doi:10.1007/s12355-019 -00729-6.

20. Mukesh Kumar Malviya, Manoj Kumar Solanki, Chang-ning Li, Reemon Htun, Rajesh Kumar Singh, Pratiksha Singh, Li-tao Yang, Yang-rui Li*. Beneficial linkages of endophytic *Burkholderia anthina* MYSP113 towards sugarcane growth promotion. Sugar Tech, 2019, 21(5):737-748.

21. P. Singh, Qi-qi Song, R.K. Singh, Hai-bi Li, M.K. Solanki, Li-tao Yang*, Yang-rui Li*. Physiological and molecular analysis of sugarcane (varieties-F134 and NCo310) during *Sporisorium scitamineum* Interaction. Sugar Tech, 2019, 21(4):631-644.

22. Jin-ju Wei, Zhi-hui Xiu, Hui-ping Ou, Jun-hui Chen, Hua-yan Jiang, Xiao-qiu Zhang, Rong-hua Zhang, Hui Zhou, Yi-yun Gui, Hai-bi Li, Yang-rui Li, Rong-zhong Yang, Dong-liang Huang, Hong-wei Tan, Xi-hui Liu*. Transcriptome profile analysis of twisted leaf disease response in susceptible sugarcane with Narenga porphyrocoma genetic background. Tropical Plant Biology,2019,12(4):1-11.

23. Xiang Li, Shan-hai Lin, Qu-yan Huang, Qiang Liang, Yi-jie Li, Li-tao Yang*, Yang-rui Li*. Advances in research of lodging and evaluation. Applied Ecology & Environmental Research, 2019, 17(3):6095-6105.

24. Xiang Li, Yi-jie Li, Qiang Liang, Shan-Hai Lin, Qu-yan Huang, Rong-zhong Yang, Li-tao Yang*, Yang-rui Li*. Evaluation of lodging resistance in sugarcane (*Saccharum* Spp. Hybrid) germplasm resources. Applied Ecology & Environmental Research, 2019, (17)3:6107-6116.

25. Xiu-peng Song, Jin-ju Wei, Xiao-qiu Zhang. Effect of sugarcane smut (*Ustilago scitaminea* Syd.) on ultrastructure and biochemical indices of sugarcane. Biomedical Journal of Scientific & Technical Research, 2019, 17 (1):12546-12551.

26. 王伦旺, 邓宇驰, 谭芳, 唐仕云, 黄海荣, 经艳, 杨荣仲. 机械化生产对桂糖47号宿根能力的影响与分析. 西南农业学报, 2019, 32(9):2163-2166.
Lun-wang Wang, Yu-chi Deng, Fang Tan, Shi-yun Tang, Hai-rong Huang, Yan Jing, Rong-zhong Yang. Effect of mechanized Production on Ratooning Ability of Guitang47. Southwest China Journal of Agricultural Sciences, 2019, 32(9): 2163-2166.

27. 王伦旺, 廖江雄, 谭芳, 邓宇驰, 唐仕云, 黄海荣, 李翔, 经艳, 杨荣仲. 甘蔗新品种"桂糖42号"对不同施氮水平的响应. 热带农业科学, 2019, 39 (5):17-20
Lun-wang Wang, Jiang-xiong Liao, fang Tan, Yu-chi Deng, Shi-yun Tang, Hai-rong Huang, Xiang Li, Yan Jing, Rong-zhong Yang. New sugarcane variety 'Guitang 42' in response to different nitrogen levels. Chinese Journal of Tropical Agriculture, 2019, 39(5):17-20.

28. 邓宇驰, 贤武, 黄杏, 黄海荣, 经艳, 王伦旺*. 种植不同甘蔗品种经济效益分析. 种子, 2019, 38(9):132-134.

Yu-chi Deng, Wu Xian, Xing Huang, Hai-rong Huang, Yan Jing, Lun-wang Wang*. Economic benefit analysis of planting different sugarcane varieties. Seed, 2019, 38(9):32-134.

29. 杨荣仲，周会，雷敬超，段维兴，唐仕云，李文教，杨翠芳，高轶静，周珊. 不同甘蔗品种收获指数研究. 中国糖料, 2019, 41(1):8-12.

    Rong-zhong Yang, Hui Zhou, Jing-chao Lei, Wei-xing Duan, Shi-yun Tang, Wen-jiao Li, Cui-fang Yang, Yi-jing Gao, Shan Zhou. The harvest index of different sugarcane varieties. Sugar Crops of China, 2019, 41(1):8-12.

30. 王泽平，林善海，梁强，李长宁，宋修鹏，刘璐，李毅杰*. 甘蔗响应梢腐病菌侵染的蛋白质组学分析. 热带作物学报, 2019, 40(5):939-946.

    Ze-ping Wang, Shan-hai Lin, Qiang Liang, Chang-ning Li, Xiu-peng Song, Lu Liu, Yi-jie Li*. Proteomic analysis of different sugarcane genotypes in response to pokkah boeng disease. Chinese Journal of Tropical Crops, 2019, 40(5): 939-946.

31. 龙盛风，李文教，韦绍龙，林善海，李毅杰，梁强，王泽平*. 不同种植模式下甘蔗梢腐病和香蕉枯萎病病害调查及其病原真菌分离鉴定. 南方农业学报, 2019, 50(2):292-298.

    Sheng-feng Long, Wen-jiao Li, Shao-long Wei, Shan-hai Lin, Yi-jie Li, Qiang Liang, Ze-ping Wang*. Disease investigation and fungi isolation on sugarcane pokkah boeng and banana vascular wilt under different cropping systems. Journal of Southern Agriculture, 2019, 50(2):292-298.

32. 雷敬超，周会，杨荣仲，高丽花，李翔，黄海荣，段维兴，经艳，王伦旺，张革民，吴杨，高轶静*. 桂糖系列甘蔗亲本开花习性及其遗传分析. 热带作物学报, 2019, 40(1):11-17.

    Jing-chao Lei, Hui Zhou, Rong-zhong Yang, Li-hua Gao, Xiang Li, Hai-rong Huang, Wei-xing Duan, Yan Jing, Lun-wang Wang, Ge-min Zhang, Yang Wu, Yi-jing Gao*. Flowering habit and heritability for GT sugarcane parents. Chinese Journal of Tropical Crops, 2019, 40(1):11-17.

33. 雷敬超，周会，高丽花，韦金菊，黄海荣，邓宇驰，李翔，李杨瑞，杨荣仲，吴杨，高轶静*. 美国新引进甘蔗亲本开花习性及其遗传分析. 西南农业学报, 2019, 32(1):25-29.

    Jing-chao Lei, Hui Zhou, Li-hua Gao, Jin-ju Wei, Hai-rong Huang, Yu-chi Deng, Xiang Li, Yang-rui Li, Rong-zhong Yang, Yang Wu, Yi-jing Gao*. Flowering habit and heritability analysis on newly introduced sugarcane parents from America. Southwest China Journal of Agricultural Sciences, 2019, 32(1): 25-29.

34. 雷敬超，周会，高丽花，杨翠芳，周珊，张保青，王泽平，杨荣仲，李翔，高轶静*. 甘蔗亚硫酸法养茎杂交影响因素分析. 种子, 2019, 38(10):104-106.

    Jing-chao Lei, Hui Zhou, Li-hua Gao, Cui-fang Yang, Shan Zhou, Bao-qing Zhang, Ze-ping Wang, Rong-zhong Yang, Xiang Li, Yi-jing Gao*. Influencing factors analysis of hybridization of sugarcane stem cultured in sulphuric acid solution. Seed, 2019, 38(10):104-106.

35. 黄玉新，张保青，高轶静，段维兴*，周珊，杨翠芳，张革民，王泽平. 广西斑茅表型性状遗传多样性分析. 热带作物学报, 2019, 40(9):1706-1712.

    Yu-xin Huang, Bao-qing Zhang, Yi-jing Gao, Wei-xing Duan*, Shan Zhou, Cui-fang Yang, Ge-min Zhang, Ze-ping Wang. Phenotypic traits and genetic diversity of *Erianthus arundinaceum* germplasm from Guangxi. Chinese Journal of Tropical Crops, 2019, 40(9):1706-1712.

36. 黄玉新, 段维兴*, 张保青, 杨翠芳, 高轶静, 周珊, 张革民, 李翔. 138份国外引进甘蔗品种(系)宿根性评价. 云南农业大学学报(自然科学), 2019, 34(4):564-570.

    Yu-xin Huang, Wei-xing Duan*, Bao-qing Zhang, Cui-fang Yang, Yi-jing Gao, Shan Zhou, Ge-min Zhang, Xiang Li. Evaluation of Ratoon Characteristics of 138 Exotic Sugarcane Germplasm. Journal of Yunnan Agricultural University (Natural Science), 2019, 34(4):564-570.

37. 周慧文, 陈荣发, 范业赓, 丘立杭, 李杨瑞, 吴建明*, 黄杏*. 不同施氮水平下甘蔗内源激素、产量和糖分的变化特征. 热带作物学报, 2019, 40(11):2142-2148.

    Hui-wen Zhou, Rong-fa Chen, Ye-geng Fan, Li-hang Qiu, Yang-rui Li, Jian-ming Wu*, Xing Huang*. The change characteristics of endogenous hormone and quality content of sugarcane under different nitrogen fertilization. Chinese Journal of Tropical Crops, 2019, 40(11):2142-2148.

38. 周慧文, 范业赓, 陈荣发, 丘立杭, 黄杏, 翁梦苓, 周忠凤, 吴建明*. 不同体积切种处理对甘蔗单芽育苗效果的影响. 广西糖业, 2019, (4):3-7.

    Hui-wen Zhou, Ye-geng Fan, Rong-fa Chen, Li-hang Qiu, Xin Huang, Meng-ling Weng, Zhong-feng Zhou, Jian-ming Wu*. Effects of the stem with single bud treated by different cutting treatments on plantlets growth of sugarcane. Guangxi Sugar Industry, 2019, (4):3-7.

39. 周慧文, 范业赓, 黄杏, 陈荣发, 杨柳, 卢星高, 吴建明*, 丘立杭*, 李杨瑞. 甘蔗健康种苗原苗提前移栽对田间繁育的影响. 中国糖料, 2019, 41(3):23-27.

    Hui-wen Zhou, Ye-geng Fan, Xing Huang, Rong-fa Chen, Liu Yang, Xing-gao Lu, Jian-ming Wu*, Li-hang Qiu*, Yang-rui Li. Effects of early transplanting on the growth and developments of virus-free sugarcane seedlings. Sugar Crops of China, 2019, 41(3):23-27.

40. 甘崇琨, 周慧文, 陈荣发, 范业赓, 丘立杭, 黄杏, 李杨瑞, 卢星高, 吴建明*. 化学调控在甘蔗生产上的研究应用. 生物技术通报, 2019, 35(2):163-170.

    Chong-kun Gan, Hui-wen Zhou, Rong-fa Chen, Ye-geng Fan, Li-hang Qiu, Xing Huang, Yang-rui Li, Xing-gao Lu, Jian-ming Wu*. Application of chemical regulating technology in sugarcane production. Biotechnology Bulletin, 2019, 35(2): 163-170.

41. 谢金兰*, 李长宁, 李毅杰, 梁强, 刘晓燕, 罗霆, 林丽, 梁阗, 何为中, 谭宏伟. 钾肥施用量对甘蔗产量、糖分积累及其抗逆性的效应研究. 中国土壤与肥料, 2019, (2):133-138.

    Jin-lan Xie*, Chang-ning Li, Yi-jie Li, Qiang Liang, Xiao-yan Liu, Ting Luo, Li Lin, Tian Liang, Wei-zhong He, Hong-wei Tan. Effects of potassium fertilizer application amount on sugarcane yield, sugar accumulation and stress resistance. Soil and Fertilizer Sciences in China, 2019, (2):133-138.

42. 范业赓, 丘立杭, 黄杏, 周慧文, 甘崇琨, 李杨瑞, 杨荣仲, 吴建明*, 陈荣发*. 甘蔗节间伸长过程赤霉素生物合成关键基因的表达及相关植物激素动态变化. 植物学报, 2019, 54(4):486-496.

Ye-geng Fan, Li-hang Qiu, Xing Huang, Hui-wen Zhou, Chong-kun Gan, Yang-rui Li, Rong-zhong Yang, Jian-ming Wu*, Rong-fa Chen*. Expression analysis of key genes in gibberellin biosynthesis and related phytohormonal dynamics during sugarcane internode elongation. Chinese Bulletin of Botany, 2019, 54(4):486-496.

43. 丘立杭,范业赓,周慧文,陈荣发,黄杏,罗含敏,杨荣仲,段维兴,刘俊仙,吴建明*. 合理密植下强分蘖甘蔗品种性状及产量分析. 热带作物学报, 2019, 40(6):1075-1082.
Li-hang Qiu, Ye-geng Fan, Hui-wen Zhou, Rong-fa Chen, Xing Huang, Han-min Luo, Rong-zhong Yang, Wei-xing Duan, Jun-xian Liu, Jian-ming Wu*. Analysis of rational close planting with agronomic traits and yield in intense tillering ability sugarcane variety. Chinese Journal of Tropical Crops, 2019, 40(6):1075- 1082.

44. 范业赓,廖洁,王天顺,丘立杭,陈荣发,黄杏,莫磊兴*,吴建明*. 镉胁迫对甘蔗抗氧化酶系统及非蛋白巯基物质的影响. 湖南农业科学, 2019, (4):23-27.
Ye-geng Fan, Jie Liao, Tian-shun Wang, Li-hang Qiu, Rong-fa Chen, Xing Huang, Lei-xing Mo*, Jian-ming Wu*. Effects of cadmium stress on antioxidant enzyme system and non-protein thiols substances in sugarcane. Hunan Agricultural Sciences, 2019, (4):23-27

45. 范业赓,丘立杭,陈荣发,周慧文,黄杏,卢星高,甘崇琨,吴建明*,李杨瑞*. 施氮水平对不同甘蔗品种产量和蔗糖分的影响. 中国糖料, 2019, 41(4):36-40.
Ye-geng Fan, Li-hang Qiu, Rong-fa Chen, Hui-wen Zhou, Xing Huang, Xing-gao Lu, Chong-kun Gan, Jian-ming Wu*, Yang-rui Li*. Effects of nitrogen application level on yield and sucrose content of different sugarcane cultivars. Sugar Crops of China, 2019, 41(4):36-40.

46. 卢星高,范业赓,丘立杭,陈荣发,黄杏,吴建明*. 广西"双高"糖料蔗基地建设现状、问题及发展建议. 热带农业科技, 2019, 42(2):51-54.
Xing-gao Lu, Ye-geng Fan, Li-hang Qiu, Rong-fa Chen, Xing Huang, Jian-ming Wu*. The current state of sugarcane base under construction and its suggestions on development in Guangxi. Tropical Agricultural Science & Technology, 2019, 42(2):51-54.

47. 范业赓,陈荣发,周慧文,丘立杭,黄杏,吴建明*,李杨瑞*. 不同植物生长调节剂浸种对甘蔗分蘖及产量性状的影响. 中国糖料, 2019, 41(2):23-27.
Ye-geng Fan, Rong-fa Chen, Hui-wen Zhou, Li-hang Qiu, Xing Huang, Jian-ming Wu*, Yang-rui Li*. Effects of seed-cane soaking by various plant growth regulators on tillering and yield traits of sugarcane. Sugar Crops of China, 2019, 41(2):23-27.

48. 卢星高,甘崇琨,范业赓,丘立杭,陈荣发,黄杏,周慧文,李杨瑞,吴建明*. 广西甘蔗良种繁育推广体系建设与发展. 中国糖料, 2019, 41(1):76-80.
Xing-gao Lu, Chong-kun Gan, Ye-geng Fan, Li-hang Qiu, Rong-fa Chen, Xing Huang, Hui-wen Zhou, Yang-rui Li, Jian-ming Wu*. Construction and development of cane setts propagation and popularizing system in Guangxi. Sugar Crops of China, 2019, 41(1):76-80.

49. 张保青,邵敏,黄玉新,黄杏,宋修鹏,陈虎,王盛,谭秦亮,杨丽涛*,李杨瑞*. 甘蔗抗坏血酸过氧化物酶基因 *ScAPX1* 的克隆和表达分析. 生物技术通报, 2019, 35(12):31-37.

Bao-qing Zhang, Min Shao, Yu-xin Huang, Xing Huang, Xiu-peng Song, Hu Chen, Sheng Wang, Qin-liang Tan, Li-tao Yang*, Yang-rui Li*. Cloning and expression analysis of peroxidase gene (*ScAPX1*) from sugarcane. Biotechnology Bulletin, 2019, 35(12):31-37.

50. 韦金菊, 宋修鹏, 魏春燕, 张小秋, 黄伟华, 颜梅新*. 甘蔗黑穗病及其防治研究进展. 广东农业科学, 2019, 46(4):81-88.

Jin-ju Wei, Xiu-peng Song, Chun-yan Wei, Xiao-qiu Zhang, Wei-hua Huang, Mei-xin Yan*. Research Progress on Sugarcane Smut and Its Control. Guangdong Agricultural Sciences, 2019, 46(4):81-88.

51. 周珊, 高轶静, 张保青, 黄玉新, 段维兴, 杨翠芳, 王泽平, 张革民*. 斑茅割手密复合体杂交利用过程野生特异基因遗传分析. 植物遗传资源学报, 2019, 20(3):718-727.

Shan Zhou, Yi-jing Gao, Bao-qing Zhang, Yu-xin Huang, Wei-xing Duan, Cui-fang Yang, Ze-ping Wang, Ge-min Zhang*. Genetic analysis of wild specific genetic loci in the hybridization process for intergeneric hybrid complex (*Erianthus arundinaceus*×*Saccharum spontaneum*). Journal of Plant Genetic Resources, 2019, 20(3):718-727.

52. 梁阗, 何为中, 谭宏伟, 高轶静, 庞天, 李德伟, 覃振强. 不同施氮水平下固氮菌肥对甘蔗的应用效果试验. 热带农业科学, 2019, 39(5):11-16.

Tian Liang, Wei-zhong He, Hong-wei Tan, Yi-jing Gao, Tian Pang, De-wei Li, Zhen-qiang Qin. Effect of azotobacter fertilizer on sugarcane under different nitrogen levels. 热带农业科学, 2019, 39(5):11-16.

53. 刘俊仙, 熊发前*, 刘菁, 罗丽, 丘立杭, 刘丽敏, 吴建明, 刘红坚, 刘欣, 卢曼曼, 何毅波, 李松*. 用于克隆及分子标记分析的甘蔗高质量基因组DNA提取方法. 分子植物育种, 2019, 17(2):545-552.

Jun-xian Liu, Fa-qian Xiong*, Jing Liu, Li Luo, Li-hang Qiu, Li-min Liu, Jian-ming Wu, Hong-jian Liu, Xin Liu, Man-man Lu, Yi-bo He, Song Li*. High quality sugarcane DNA extraction methods for cloning and molecular marker analysis. Molecular Plant Breeding, 2019, 17(2):545-552.

54. 魏吉利, 潘雪红*, 黄诚华, 商显坤, 林善海. 温度对甘蔗条螟生长发育和繁殖的影响. 植物保护学报, 2019, 46(6):1277-1283.

Ji-li Wei, Xue-hong Pan*, Cheng-hua Huang, Xian-kun Shang, Shan-hai Lin. Effects of temperature on the development and reproduction of spotted borer *Chilo sacchariphagus* (Lepidoptera: Pyralidae). Journal of Plant Protection, 2019, 46(6):1277-1283.

55. 张小秋, 宋修鹏, 梁永检, 宋奇琦, 覃振强, 李杨瑞*, 吴建明*. 植保无人机在蔗田化学除草上的应用效果. 中国糖料, 2020, 42(1):61-65.

Xiao-qiu Zhang, Xiu-peng Song, Yong-jian Liang, Qi-qi Song, Zhen-qiang Qin, Yang-rui Li*, Jian-ming Wu*. Application effects of unmanned aerial vehicle on chemical weed control in sugarcane field. Sugar Crops of China, 2020, 42(1):61-65.

56. 经艳, 周会, 刘昔辉, 谭芳, 张小秋, 张荣华, 宋修鹏, 李杨瑞, 颜梅新, 雷敬超, 覃振强,罗亚伟，李冬梅，韦金菊*. 桂糖甘蔗新品系黑穗病抗性鉴定及结果分析. 热带作物学报，2020, 41(2):333-338.

Yan Jing, Hui Zhou, Xi-hui Liu, Fang Tan, Xiao-qiu Zhang, Rong-hua Zhang, Xiu-peng Song, Yang-rui Li, Mei-xin Yan, Jing-chao Lei, Zhen-qiang Qin, Ya-wei Luo, Dong-mei Li, Jin-ju Wei*. Smut resistant identification and analysis of new sugarcane clones of Guitang. Chinese Journal of Tropical Crops, 2020, 41(2):333-338.

57. 王凡伟，肖冬，李杨瑞*，何龙飞*，王爱勤*. 甘蔗乙烯信号转导途径关键基因 *CTR1* 的克隆与表达分析. 分子植物育种，2020, 18(3):827-833.

Fan-wei Wang, Dong Xiao, Yang-rui Li*, Long-fei He*, Ai-qin Wang*. Cloning and expression analysis of the gene of a key factor *CTR1* in sugarcane ethylene signal transduction pathway. Molecular Plant Breeding, 2020, 18(3):827-833.

58. 刘昔辉，桂意云，张荣华，李海碧，韦金菊，周会，杨荣仲，张小秋，李杨瑞*. 河八王杂交种 F_1、BC_1 及其亲本 DNA 甲基化水平和模式变化. 中国农业大学学报，2019, 24(9):37-46.

Xi-hui Liu, Yi-yun Gui, Rong-hua Zhang, Hai-bi Li, Jin-ju Wei, Hui Zhou, Rong-zhong Yang, Xiao-qiu Zhang, Yang-rui Li*. DNA methylation levels and genetic patterns in *Narenga porphyrocoma* (Hance) hybrids F_1, BC_1 and their parental inbreds. Journal of China Agricultural University, 2019, 24(9):37-46.

59. 廖芬，桂杰，杨柳*，李强，Muhammad Anas，李杨瑞*. 施用生物炭对甘蔗土壤化学性质及氮损失的影响. 广西糖业，2019, (3):36-42.

Fen Liao, Jie Gui, Liu Yang*, Anas Muhammad, Yang-rui Li*. The effect of application biochar on soil chemical property and nitrogen loss of sugarcane. Guangxi Sugar Industry, 2019, (3):36-42.

60. 廖芬，杨柳*，李强，Muhammad Anas，薛进军，黄东亮，李杨瑞*. 不同生物质来源生物炭品质的因子分析与综合评价. 华南农业大学学报，2019, 40 (3):29-37.

Fen Liao, Liu Yang*, Qiang Li, Anas Muhammad, Jin-Jun Xue, Dong-liang Huang, Yang-rui Li*. Factor analysis and comprehensive evaluation for quality of biochar derived from different biomass. Journal of South China Agricultural University, 2019, 40(3):29-37.

61. 刘昔辉，张荣华，李海碧，张革民，周会，桂意云，韦金菊，杨荣仲，宋焕忠*，李杨瑞*. 甘蔗与河八王杂交后代染色体传递及河八王血缘在后代中的遗传. 分子植物育种，2019, 17(2):411-418.

Xi-hui Liu, Rong-hua Zhang, Hai-bi Li, Ge-min Zhang, Hui Zhou, Yi-yun Gui, Jin-ju Wei, Rong-zhong Yang, Huan-zhong Song*, Yang-rui Li*. Genetic analysis of the hybrid progeny of *Saccharum officinarum* L. and *Narenga porphyrocoma* (Hance) Bor. Molecular Plant Breeding, 2019, 17(2):411-418.

62. 李海碧，桂意云，张荣华，韦金菊，杨荣仲，张小秋，李杨瑞，周会*，刘昔辉*. 甘蔗抗旱性及抗旱育种研究进展. 分子植物育种，2019, 17(10):3406-3415.

Hai-bi Li, Yi-yun Gui, Rong-hua Zhang, Jin-ju Wei, Rong-zhong Yang, Xiao-qiu Zhang, Yang-rui Li, Hui Zhou*, Xi-hui Liu*. Research progress on drought resistance and drought-resistant breeding of sugarcane. Molecular Plant Breeding, 2019, 17(10):3406-3415.

63. 韦金菊, 宋修鹏, 张荣华, 张小秋, 覃振强, 魏春燕, 桂意云, 周会, 谭宏伟, 黄东亮, 李海碧, 吴杨, 李杨瑞*, 刘昔辉*. 美国和越南引进甘蔗新种质隔离检疫及农艺性状评价. 分子植物育种, 2019, 17(14):4708-4716.

    Jin-ju Wei, Xiu-peng Song, Rong-hua Zhang, Xiao-qiu Zhang, Zhen-qiang Qin, Chun-yan Wei, Yi-yun Gui, Hui Zhou, Hong-wei Tan, Dong-liang Huang, Hai-bi Li, Yang Wu, Yang-rui Li*, Xi-hui Liu*. Assessment of agronomic traits and quarantine disease of newly introduced sugarcane germplasm from USA and Vietnam. Molecular Plant Breeding, 2019, 17(14):4708-4716.

64. 宋奇琦, Pratiksha Singh, Rajesh, Kumar Singh, 宋修鹏, 李海碧, 农友业, 杨丽涛*, 李杨瑞*. 基于iTRAQ技术的甘蔗受黑穗病菌侵染蛋白组分析. 作物学报, 2019, 45(1):55-69.

    Qi-qi Song, Pratiksha Singh, Rajesh Kumar Singh, Xiu-peng Song, Hai-bi Li, You-ye Nong, Li-tao Yang*, Yang-rui Li*. Proteomic analysis of sugarcane *Sporisorium scitamineum* interaction based on iTRAQ technique. Acta Agronomica Sinica, 2019, 45(1):55-69.

3.2 发表国际会议论文 International Conference Papers

1. Wei-zhong He, Li-min Liu, Hong-jian Liu, Tian Liang, Meng-ling Weng, Song Li. Factors affecting the survival of sugarcane micro-shoots during photoautotrophic rooting. Proceedings of the International Society of Sugar Cane Technologists, volume 30, 48, 2019.

2. Yu-chi Deng, Lun-wang Wang*, Yan Jing, Wu Xian, Hai-rong Huang, Fang Tan. Analysis on the high and sable yield character and adaptability of sugarcane variety Guitang 44. National Congress of Plant Biology, 227, 2019.

3. Liu Yang, Fen Liao, Muhammad Anas, Qiang Li, Li-shun Peng, Dong-liang Huang, Yang-rui Li. Screening of sugarcane for high nitrogen-use efficiency at the seedling stage. Proceedings of the International Society of Sugar Cane Technologists, volume 30, 1696-1702, 2019.

4. Shan-hai Lin, Yi-jie Li, Ya-wei Luo, Xiang Li, Yang-rui Li, Wei-zan Wang, Wei-zhong He. Incidence of pokkah boeng in a disease-screening field test in China. Proceedings of the International Society of Sugar Cane Technologists, volume 30, 1332-1335, 2019.

5. De-wei Li, Zhen-qiang Qin, Ya-wei Luo, Xiu-peng Song, Chun-yan Wei, H.Z. Ding, C. Lin. Effects of the host plants *Zizania latifolia* (Manchurian wild rice) and sweet corn on growth and development of the sugarcane shoot borer *Chilo infuscatellus* Snellen. Proceedings of the International Society of Sugar Cane Technologists, volume 30, 330-334, 2019.

6. Zhen-qiang Qin, De-wei Li, Xue-hong Pan, Xiu-peng Song, Chun-yan Wei, Jin-ju Wei, Ya-wei Luo. Effects of exposure time and breeding generations of *Trichogramma chilonis* field populations on parasitism of *Corcyra cephalonica* eggs. Proceedings of the International Society of Sugar Cane Technologists, volume 30, 327-329, 2019.

7. Xian-kun Shang, Cheng-hua Huang, Ji-li Wei, Xue-hong Pan, Shan-hai Lin. Soil pH and organic matter levels in the area of occurrence of the whitegrub *Alissonotum impressicolle* Arrow (Coleoptera: Dynastidae). Proceedings of the International Society of Sugar Cane Technologists, volume 30, 36-40, 2019.

8. Jin-ju Wei, Zhen-qiang Qin, Xi-hui Liu, Xiu-peng Song, Xiao-qiu Zhang, Hai-bi Li, Chun-yan Wei, Lu Liu. Sugarcane ratoon stunting disease in China: review of the current situation and disease control. Proceedings of the International Society of Sugar Cane Technologists, volume 30, 1773-1776, 2019.

9. Xiao-qiu Zhang, Xi-hui Liu, Li-tao Yang, Yang-rui Li. Transcriptome analysis reveals the genes responding to *Leifsonia xyli* subsp. *xyli* infection in sugarcane, Proceedings of the International Society of Sugar Cane Technologists, volume 30, 1288-1295, 2019.

10. Yang-rui Li, Zhen Wang, MK Solanki, Z.X. Yu, Qian An, Deng-feng Dong, Li-tao Yang. *Streptomyces chartreusis* strain WZS021 stimulates the drought tolerance of sugarcane. Proceedings of the International Society of Sugar Cane Technologists, volume 30, 1314–1323, 2019.

11. Yang-rui Li, Kai Zhu, Dan Yuan, Yong-xiu Xing, You-ye Nong, Li-tao Yang. Effects of different nitrogen levels on plant growth and expression of glutamine synthetase *SCGS1* in sugarcane. Proceedings of the International Society of Sugar Cane Technologists, volume 30, 1459-1468, 2019.

4 合作与交流
COOPERATION AND EXCHANGE

4.1 实验室人员参加国内外学术交流记录
Important Academic Exchange Activities

4.1.1 参加国外学术交流记录 International Academic Exchange Activities

（1）2019年2月15—20日，"绿色环保技术助推蔗糖产业可持续发展"国际会议（Sugarcon-2019）在勒克瑙市印度甘蔗研究所召开。实验室李杨瑞、Prakash Lakshmanan、黄东亮、王维赞、杨丽涛、陈赶林、刘昔辉、宋修鹏等一行8人应邀参加本次会议。李杨瑞和Prakash Lakshmanan 分别作为Plenary Session和Sugar Crops Improvement, Breeding and Biotechnology Section的主席，并分别在Plenary Session作"Green Technology Initiatives in Chinese Sugar Industry"和"Integration of phenomics and genomics for sugarcane variety improvement"大会主题报告。李杨瑞、杨丽涛分别在"Green Technologies to Enhance Sugar Productivity Section"作"Effect of Different Nitrogen Levels on Plant Growth and Glutamine Synthase Genes $scGS1$ Expression in Sugarcane"、"An Overview of Physio-biochemical and Molecular Work on N-fixation in Sugarcane in China"报告，陈赶林在"Sugar Production, processing, value addition and energy management in sugar industry Section"作"Effect of sugarcane vinegar on lipid metabolism and redox state in high-lipid-diet rats"报告。李杨瑞在会上获SSRP授予"全球糖业贡献奖"。

On February 15-20, 2019, the International Conference on "Green Environmental Protection Technology Promoting the Sustainable Development of the Sugar Industry" (Sugarcon-2019) was held in Lucknow, India. Yang-rui Li, Prakash Lakshmanan, Dong-liang Huang, Wei-zan Wang, Li-tao Yang, Gan-Lin Chen, Xi-hui Liu, Xiu-peng Song attended this conference. Drs. Yang-rui Li and Prakash Lakshmanan were the chairpersons of "Plenary Session" and Sugar Crops Improvement, Breeding and Biotechnology Section, respectively, and they delivered the presentations entitled "Green technology initiatives in Chinese sugar industry" and "Integration of phenomics and genomics for sugarcane variety improvement" in the Plenary Session. Profs. Yang-rui Li and Li-tao Yang gave presentations entitled "Effect of different nitrogen levels on plant growth and glutamine synthase genes $scGS1$ expression in sugarcane" and "An overview of physio-biochemical and molecular work on n-fixation in sugarcane in China" in "Green Technologies to Enhance Sugar Productivity Section", and Prof. Gan-lin Chen gave the presentations entitled "Sugar production, processing, value addition and energy management in sugar industry Section" and "Effect of sugarcane vinegar on lipid metabolism and redox state in high-lipid-diet rats". Prof. In the conference, Prof. Yang-rui Li was awarded for "Dedication and Services to Global Sugar Industry" by Society for Sugar Research Promotion.

（2）国际甘蔗技师协会（ISSCT）第30届大会（XXX ISSCT Congress）"于2019年9月2日至9月5日在阿根廷土库曼隆重召开。应大会组委会的邀请，我院甘蔗专家李杨瑞、Prakash Lakshmanan、黄东亮、刘昔辉、宋修鹏、何为中、王伦旺、梁阗、李德伟、庞天等一行11人参加了大会。本实验室人员共发表了4篇全文论文、10篇墙报论文。李杨瑞教授作了"Streptomyces chartreusis strain WZS021 stimulates the drought tolerance of sugarcane plant"的口头报告，宋修鹏博士作了"Transcriptome analysis reveals the genes responding to *Leifsonia xyli* subsp. *xyli* infection in sugarcane"。会议期间，本实验室人员还与法国Visacane参会人员一起讨论关于2020年甘蔗种质资源交换的具体事宜。

From September 2 to September 5, 2019, the XXX ISSCT Congress of the International Society of Sugar Cane Technologists (ISSCT) was held in Tucuman, Argentina. Yang-rui Li, Prakash Lakshmanan, Dong-liang Huang, Xi-hui Liu, Xiu-peng Song, Wei-zhong He, Lun-wang Wang, Tian Liang, De-wei Li and Tian Pang attended the conference. The researchers of the laboratory published 4 full peer reviewed papers and 10 poster papers in the Proceedings of ISSCT. Prof. Yang-rui Li gave an oral presentation entitled "Streptomyces chartreusis strain WZS021 stimulates the drought tolerance of sugarcane plant", and Dr. Xiu-peng Song gave an oral presentation entitled "Transcriptome analysis reveals the genes responding to *Leifsonia xyli* subsp. *xyli* infection in sugarcane". The delegates from the laboratory also met and discussed with the attendees from Visacane, France about the details of sugarcane germplasm exchange in 2020.

（3）2019年11月25日至12月2日，实验室李鸣、李松研究员访问美国农业部运河点甘蔗研究所。

From November 25th to December 2nd, 2019, Profs. Ming Li and Song Li visited the USDA-ARS Sugarcane Field Station, Canal Point, USDA.

（4）李杨瑞、Prakash Lakshmanan、黄东亮、刘昔辉、张保青、宋修鹏等应邀出席2019年11月12—15日在昆明举办的"世界主要蔗糖生产国甘蔗科技合作交流研讨会（Symposium on Science and Technology of Sugarcane Cooperation and Exchange of Major Sugar Producing Countries in the World）"。李杨瑞和Prakash Lakshmanan受邀分别作"Development Strategy of Sugarcane Industry in China"和"Constraints and Opportunities for Surgarcane Variety and Crop Improvement in China"的学术报告。

Yang-rui Li, Prakash Lakshmanan, Dong-liang Huang, Xi-hui Liu, Bao-qing Zhang and Xiu-peng Song attended "Symposium on Science and Technology of Sugarcane Cooperation and Exchange of Major Sugar Producing Countries in the World" held in Kunming, 12-15 November 2019. Yang-rui

Li and Prakash Lakshmanan were invited delivering the presentation entitled "Development Strategy of Sugarcane Industry in China" and "Constraints and Opportunities for Sugarcane Variety and Crop Improvement in China", respectively, in the symoposium.

（5）2019年12月8—14日，李杨瑞教授参与中国赴印度调研糖市代表团，分别访问新德里（New Delhi）、马邦（Maharashtra）和北方邦（Ultra Pradesh），请著名糖业研究机构的甘蔗专家（印度农业科学研究所的G.P. Rao博士、北方邦甘蔗研究局S.K. Pathak博士，马邦Vasantdada糖业研究所多位研究人员）及食糖贸易行（Sucden和Alvean）的专家介绍印度近来印度糖业生产、甘蔗品种、食糖市场、贮备、市场营销情况和未来发展趋势等情况，先后访问了两个代表性糖厂，即马邦的Shri Vighnahar Sahakari Sakhar Karkhana Ltd.和北方邦的Simbhaoli Sugars及其蔗区。

On December 8-14, 2019, Prof. Yang-rui Li joined the Chinese delegation for investigating the sugar marketing in India, and visited New Delhi, Maharashtra and Ultra Pradesh states of India. They attended the presentations delivered by Dr. G.P. Rao, chief scientist of Indian Institute of Agricultural Science, Dr. S.K. Pathak, extension officer of U.P. Council of Sugarcane Research, and several other experts from Vasantdada Sugar Institute, sugar marketing companies, Sucden and Alvean, about sugar production, sugarcane varieties, and sugar market, storage, marketing management and future development trend in India. The delegation also visited two representative sugar mills, Shri Vighnahar Sahakari Sakhar Karkhana Ltd. in Maharashtra and Simbhaoli Sugars in Ultra Pradesh, and their sugarcane growing areas.

4.1.2 参加国内学术交流记录 Domestic Academic Exchange Activities

（1）2019年1月10—12日，谭宏伟研究员参加了在云南省澜沧县举行国家重点研发计划"特色经济作物化肥农药减施技术集成研究与示范"项目2018年度总结交流会，并作"甘蔗化肥农药减施增效技术集成研究"报告。

On January 10-12, 2019, Prof. Hong-wei Tan participated in the 2018 Annual Meeting of the National Key R & D Program project "Research and Demonstration on Integrated Technologies for Reduction of Chemical Fertilizer and Pesticide Application in Special Cash Crops" in Lancang County, Yunnan Province. He delivered a speech entitled "Research on the integrated technology for increasing benefits by reducing application of chemical fertilizers and pesticides in sugarcane".

（2）3月1日，谭宏伟、Prakash Lakshmanan、黄东亮、刘昔辉赴中国农业大学拜访中国工程院院士、中国农业大学张福锁教授。

On March 1, Hong-wei Tan, Prakash Lakshmanan, Dong-liang Huang, and Xi-hui Liu visited the Academician of Chinese Academy of Engineering, Professor Fu-suo Zhang in China Agricultural University.

（3）2019年4月13—15日，李杨瑞、王维赞、宋修鹏、张小秋一行4人应邀赴中国科学院近代物理研究所兰州重离子加速器重点实验室进行交流。

On April 13-15, 2019, Yang-rui Li, Wei-zan Wang, Xiu-peng Song and Xiao-qiu Zhang visited the Institute of Modern Physics, Chinese Academy of Sciences.

（4）2019年4月12日，"甘蔗产业振兴发展大会"在来宾市召开，李杨瑞教授和王维赞研究员应邀参会分别作"甘蔗全面放开后的发展新格局"和"'双高'基地糖料蔗机械化生产技术"的报告。

On April 12, "Sugarcane Industry Vitalization and Development Workshop" was held in Laibin City, Profs. Yang-rui Li and Wei-zan Wang were invited to attend the workshop, and they gave speeches entitled "The new development pattern of sugarcane industry after complete opening-up" and "Machanization technology of sugarcane production in the 'double-high (high yield and high sugar)' bases".

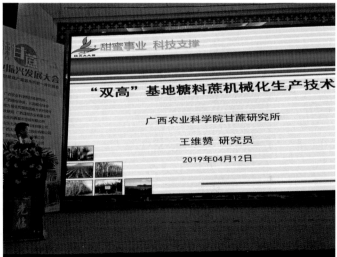

（5）2019年5月14—17日，以"推进现代特色农业高质量发展"为主题的九三学社中央第二十一次科学座谈会在南宁召开。谭宏伟以"高效甘蔗生产服务蔗糖业转型升级"为题作交流发言。

On May 14-17, 2019, the 21st Science Symposium of the Central Committee of Jiu San Society with the theme of "Promoting the High-Quality Development of Modern Agriculture with Special Features" was held in Nanning. Prof. Hong-wei Tan gave a speech on "Efficient sugarcane production service transformation and upgrade of the sugar industry".

（6）2019年5月24—26日，以"加快糖业转型升级，共享丝路合作商机"为主题的2019年中国糖业博览会暨世界糖业研讨会在广西南宁国际会展中心举办。谭宏伟研究员参加糖博会分论坛"高产高糖糖料发展论坛"会议，并作为演讲嘉宾作了题为"高效甘蔗生产服务蔗糖业转型升级"的主题演讲。

From May 24 to 26, 2019, the 2019 China Sugar Expo and World Sugar Workshop with the theme of "Accelerating the transformation and upgrading of the sugar industry and sharing the business opportunities of the Silk Road" was held at the Nanning International Convention Center, Guangxi. Prof. Hong-wei Tan participated in the Sugar Expo sub-forum on "High-Yield and High-Sugar Crops Development Forum", and gave a keynote speech entitled "Transformation and Upgrade of Efficient Sugarcane Production Services for the Sugar Industry".

（7）2019年6月18日，谭宏伟带队参加第17届中国海峡项目成果交易会，福建省农科院签订战略合作协议。

On June 18, 2019, Prof. Hong-wei Tan led a team to participate in the 17th China Strait Project Achievement Fair, and signed an agreement on strategic cooperation with Fujian Academy of Agricultural Sciences.

（8）"广西甘蔗学会第十一届代表大会暨2019年学术交流会"7月10—13日在百色田东县召开，各有关单位代表200余人参加了会议，百色市委常委、副市长李联成在开幕式上致辞。本实验室主任、广西甘蔗学会理事长李杨瑞教授代表上一届理事会做工作总结报告。李杨瑞、吴建明、Prakash Lakshmanan、陈赶林、王维赞、颜梅新等多位专家在会上作了专题报告。与会代表还参观了百色市农业科学研究所甘蔗新品种新技术试验、示范基地、广西百色国家农业科技园区、芒果智慧庄园等。本次会议顺利完成了第十一届理事会换届选举工作，本实验室主任李杨瑞教授继续当选新一届广西甘蔗学会理事长，谭宏伟和吴建明研究员当选副理事长，王维赞研究员当选秘书长，张保青博士和宋修鹏博士当选副秘书长，黄东亮当选为监事会主席。

The 11th Guangxi Sugarcane Society Congress and 2019 Annual Meeting was held in Tiandong County, Baise City, Guangxi on July 10-13, 2019. More than 200 delegates from different institutions attended the meeting, Mr. Lian-cheng Li, Vice Mayor of Baise City addressed the opening ceremony. Prof. Yang-rui Li, Director of the laboratory and President of Guangxi Sugarcane Society, made a summary report on behalf of the last Council. Yang-rui Li, Jian-ming Wu, Prakash Lakshmanan, Gan-lin Chen, Wei-zan Wang and Mei-xin Yan delivered specific presentations in the meeting. The delegates visited the field experiments and demonstrations for new sugarcane varieties and cultivation technologies set by Baise Institute of Agricultural Sciences, Guangxi Baise National Agricultural Science and Technology Park, Mango Wisdom Manor, etc. This congress successfully completed the election of the 11th Council. Prof. Yang-rui Li was re-elected as the President, Prof. Hong-wei Tan and Dr. Jian-ming Wu as vice-presidents, Wei-zan Wang as secretary-general, Drs. Bao-qing Zhang and Xiu-peng Song as vice secretary-general, and Dong-liang Huang as chairman of the supervisory board.

广西甘蔗学会第十一届代表大会暨２０１９年学术交流会集体合影

（9）2019年7月20—21日，王维赞、庞天、覃文宪一行3人到农业农村部南京农业机械化研究所交流。

On July 20-21, 2019, Wei-zan Wang, Tian Pang and Wen-xian Qin visited Nanjing Agricultural Mechanization Research Institute of the Ministry of Agriculture and Rural Affairs.

（10）2019年7月27—31日，谭宏伟研究员带领国家糖料产业技术体系广西岗位专家、试验站站长及其团队成员一行20多人赴福建农林大学国家甘蔗工程技术研究中心、福建省农业科学院亚热带农业研究所和广东省科学院生物工程研究所考察交流，参加考察。

From July 27 to 31, 2019, Prof. Hong-wei Tan led a team of more than 20 people including the experts, heads of experimental stations and members from the National Sugar Crops Industry Technology System in Guangxi visited the National Sugarcane Engineering Technology Research Center of Fujian Agriculture and Forestry University, Subtropical Agriculture Research Institute of Fujian Academy of Agricultural Sciences and Bioengineering Research Institute of Guangdong Academy of Sciences.

（11）2019年10月27—30日，实验室主任李杨瑞教授带队参加了在浙江省杭州召开的"中国作物学会第十一届会员代表大会暨2019年学术年会"。大会主题为"科技创新与扶贫攻坚"。

On October 27-30, 2019, Professor Yang-rui Li, the Director of the key laboratory, led a team to participate in the "Eleventh Membership Congress of the Chinese Crop Society and the 2019 Annual Academic Conference" held in Hangzhou, Zhejiang Province. The theme of the conference was "Technological Innovation and Poverty Alleviation".

（12）2019年10月30日，李杨瑞教授应邀在南宁市召开的"广西数字农业峰会"上做"用数字改造'三农'"的学术报告。

On October 30, 2019, Professor Yang-rui Li was invited to deliver a presentation entitled "Digital transformation of agriculture, rural areas and farmers" in "Guangxi Digital Agriculture Summit" held in Nanning.

(13) 2019年11月24—27日，实验室李杨瑞、刘昔辉、吴凯朝、一行6人参加在福建福州召开的2019年全国热带作物遗传育种学术研讨会，刘昔辉研究员在会上作了"甘蔗近缘属河八王创新利用及优异基因挖掘"的学术报告。

On November 24-27, 2019, Prof Yang-rui Li and a team of 6 people participated in the 2019 National Tropical Crops Genetics and Breeding Academic Conference held in Fuzhou, Fujian. Xi-hui Liu gave a speech on "Innovative use of *Narenga* and its elite gene mining".

4.2 国内外专家来实验室进行学术交流 Academic Exchange Activities with Foreign and Domestic Visitors in Laboratory

(1) 2019年1月21日，中国科学院华南植物园马国华研究员到实验室考察交流。

Prof. Guo-hua Ma from the South China Botanical Garden of Chinese Academy of Science visited the laboratory on January 21, 2019.

(2) 2019年3月1日，国家玉米产业技术体系病虫草害防控研究室主任、中国农业科学院植物保护研究所农业昆虫研究室主任王振营研究员到实验室访问交流。

Prof. Zhen-ying Wang, Director of Division of Disease and Pest Control Research in the National Corn Industry Technology System, and Director of Division of Agricultural Insect Research, Institute of Plant Protection, Chinese Academy of Agricultural Sciences, visited the laboratory on March 1, 2019.

（3）2019年3月21日，农业农村部南京农业机械化研究所、中国农业科学院种植机械创新团队首席专家张文毅研究员，国家麻类产业技术体系机械化研究室主任及种植与收获机械化岗位专家李显旺研究员等农业机械专家一行4人应邀到实验室及广西蔗区考察调研。

Prof. Wen-yi Zhang, chief expert of the Innovative Team of Planting Machinery, Chinese Academy of Agricultural Sciences, Nanjing Agricultural Mechanization Research Institute, Ministry of Agriculture and Rural Affairs, Xian-wang Li, Director of the Mechanization Research Office and Scientist of Post of Planting and Harvesting Mechanization in the National Hemp Industry Technology System, led a group of 4 members to visit the laboratory and Guangxi sugarcane area on March 21, 2019.

（4）2019年3月28日，谭宏伟副院长陪同西班牙瑞萨集团农田灌溉专家一行2人到访问实验室并到崇左考察有机甘蔗生产。

Vice President Hong-wei Tan and the farmland irrigation experts from Spanish Renesas Group visited the laboratory and the organic sugarcane production in Chongzhuo city on March 28, 2019.

(5) 2019年4月17日，中国工程院院士张福锁教授与邓国富院长到实验室与广西、云南、广东甘蔗科技人员进行座谈。

Prof. Fu-suo Zhang, academician of Chinese Academy of Engineering, and President Guo-fu Deng visited the laboratory and held a seminar with sugarcane scientists from Guangxi, Yunnan, and Guangdong provinces on April 17, 2019.

(6) 2019年4月19日，内蒙古农牧业科学院副院长、国家糖料产业技术体系首席科学家白晨到实验室调研。

Chen Bai, Vice President of Inner Mongolia Academy of Agriculture and Animal Husbandry Sciences and chief expert of National Sugar Crops Industry Technology System, visited the laboratory on April 19, 2019.

(7) 2019年5月27日，美国糖业公司农业研发部首席农艺师胡承健先生来广西农业科学院甘蔗研究所作"现代农业及农业航空"学术报告。

Mr. Cheng-jian Hu, the chief agronomist of Division of Agricultural Research and Development in the U.S. Sugar Corporation, visited the laboratory and delivered a presentation on "Modern Agriculture and Agricultural Aviation" on May 27, 2019.

(8) 2019年6月14日，中国科学院微生物研究所仲乃琴研究员一行到实验室进行学术交流。

Prof. Nai-qin Zhong from Institute of Microbiology of Chinese Academy of Science visited the laboratory on June 14, 2019.

(9) 2019年11月27日，印度农业科学院科学家、国际著名甘蔗植物病理专家G.P. Rao博士到实验室访问，并作"印度甘蔗产业现状"学术报告。

Dr. G.P. Rao, an internationally renowned expert on sugarcane pathology from Indian Agricultural Research Institute visited the laboratory and gave a lecture on "The Status of the Indian Sugarcane Industry" on November 27, 2019.

(10) 2019年12月13日，中国热带农业科学院甘蔗研究中心伍苏然博士和广东省科学院生物工程研究所（广州甘蔗糖业研究所）副所长安玉兴研究员一行2人到实验室考察交流。

Dr. Su-ran Wu from the Sugarcane Research Center of Chinese Academy of Tropical Agricultural Sciences, and Prof. Yu-xing An, Deputy Director of Institute of Bioengineering of Guangdong Academy of Science (Guangzhou Sugarcane and Canesugar Research Institute), visited the laboratory on December 13, 2019.

5 甘蔗科研进展
PROGRESS IN SUGARCANE RESEARCH

5.1 甘蔗种质创新与育种
Sugarcane Germplasm Innovation and Breeding

5.1.1 甘蔗种质创新 Sugarcane Germplasm Innovation

桂糖系列甘蔗亲本开花习性及其遗传分析

为明确桂糖系列甘蔗亲本的开花习性，2011—2018年连续7个杂交季节对中国农业科学院甘蔗研究中心海南甘蔗杂交基地的25个桂糖系列亲本的始花期、抽穗率和花粉染色率进行了调查和分析（图1）。结果表明，甘蔗亲本始花期、抽穗率和花粉染色率的广义遗传力分别为0.950 2、0.963 1和0.923 6，均属遗传力较高的性状；始花期最早的亲本是桂糖28号，最迟的是桂糖11号，2011—2012年杂交季节各亲本的始花期普遍最早，2017—2018年杂交季各亲本的始花期普遍最晚；抽穗率低的亲本是桂糖11号、桂糖21号、桂糖32号、桂糖43号和桂糖47号，这5个亲本属于难开花亲本，其余20个亲本均属于易开花亲本；花粉染色率低的亲本是桂糖24号、桂糖34号、桂糖45号和桂糖48号，这4个亲本的花粉染色率历年均低于30%，杂交时只能作为母本，桂糖28号、桂糖41号和桂糖44号的花粉染色率最高。以上结果为桂糖系列甘蔗亲本的杂交育种工作提供重要的依据。

（雷敬超，周会，杨荣仲，高丽花，李翔，黄海荣，段维兴，经艳，王伦旺，张革民，吴杨，高轶静*）

Flowering habit and its heritability in GT sugarcane parental clones

In order to understand the flowering habits for GT sugarcane parents, the initial flowering time, heading rate and pollen viability rate, 25 GT parents were investigated during 2011-2018 crossing seasons at Hainan Sugarcane Crossing Station of Sugarcane Research Center, Chinese Academy of Agricultural Sciences

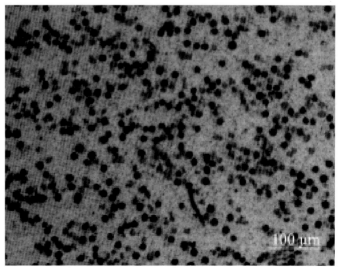

图1 2017—2018年杂交季节桂糖28号花粉镜检结果

Fig.1 Microscopic pollen examination result of GT28 in 2017—2018 crossing season

(Fig.1). The results showed that broad-sense heritability of the initial flowering time, the heading rate and the pollen viability rate were 0.950 2, 0.963 1 and 0.923 6, respectively, indicating that the three characters were of high heritable. The parent with the earliest flowering time was GT28 and the one that flowered last in the trial was GT11. Twenty-five parents showed early flowering phenotype in 2011—2012 and in 2017—2018 crossing seasons. The parents with shy flowering phenotype were GT11, GT21, GT32, GT43 and GT47, and the rest 20 clones were easy flowering parents. The parents with low pollen viability were GT24, GT34, GT45 and GT48, which are good to be used as female parents. The parents with high pollen viability were GT28, GT41 and GT44. These results provided a useful reference for crossing work with GT sugarcane parents.

(Jing-chao Lei, Hui Zhou, Rong-zhong Yang, Li-hua Gao, Xiang Li, Hai-rong Huang, Wei-Xing Duan, Yan Jin, Lun-wang Wang, Ge-min Zhang, Yang Wu, Yi-jing Gao*)

新引进美国CP甘蔗亲本开花习性及其遗传分析

明确12个美国新引进甘蔗亲本的开花习性，为甘蔗育种的亲本选配提供参考依据（图2）。于2013—2014年至2017—2018年连续5年杂交季对该亲本的始花期、抽穗率和花粉染色率进行调查和分析。甘蔗亲本始花期、抽穗率和花粉染色率的广义遗传力分别为0.951 8、0.796 1和0.938 4；始花期最早的亲本是CP07-1618和CP09-4448，最迟的是CP07-2137，2015—2016杂交季各亲本的始花期普遍最晚，其他年份杂交季各亲本的始花期差异不明显；抽穗率低的亲本是CP01-2390、CP07-2137和CP09-4707，均属于难开花亲本，其余9个亲本均属于易开花亲本；花粉染色率低的亲本是CP00-1630、CP06-3458、CP07-1618和CP07-2137，其花粉染色率历年均不超过30%，杂交时只能作为母本，CP06-2422和CP09-4448的花粉染色率最高。甘蔗亲本始花期、抽穗率和花粉染色率的广义遗传力均属遗传力较高的性状，可在杂交组合配制阶段对该亲本进行较严格的选择。

（雷敬超，周会，高丽花，韦金菊，黄海荣，邓宇驰，李翔，李杨瑞，杨荣仲，吴杨，高铁静*）

Flowering habit and heritability analysis of newly introduced CP sugarcane parents from U.S.A.

To provide reference basis for parents selection and matching of sugarcane breeding, the flowering habits of 12 newly introduced CP sugarcane parents from USDA-ARS Sugarcane Field Station, Canal Point, Florida, U.S.A. were determined (Fig. 2). The initial flowering time, heading rate and pollen viability were investigated and analyzed for continuous 5 crossing seasons from 2013/2014 to 2017/2018. Broad-sense heritability of the initial flowering time, the heading rate and the pollen viability were 0.951 8, 0.796 1 and 0.938 4, respectively. The parents with the earliest first flowering time were CP07-1618 and CP09-4448, and the parent with the latest first flowering time was CP07-2137. This flowering behaviour was observed in 2015/2016 crossing season. There was no significant difference in initial flowering time among the parents in other crossing seasons. The parents with low heading rate were CP01-2390, CP07-2137 and CP09-4707, belonging to shy flowering parents, and the other clones were easy flowering parents. The parents with low pollen viability were CP00-1630, CP06-3458, CP07-

1618, CP07-2137, which could only be used as female parents. Theirpollen viability rates were no more than 30% in all crossing seasons. The parents with high pollen viability rate were CP06-2422 and CP09-4448. The broad-sense heritability characters of the initial flowering time, heading rate and pollen viability of sugarcane parents were high heritability, so the 12 parents could be selected for appropriate crossing combinations.

(Jing-chao Lei, Hui Zhou, Li-hua Gao, Jin-ju Wei, Hai-rong Huang, Yu-chi Deng, Xiang Li, Yang-rui Li, Rong-zhong Yang, Yang Wu, Yi-jing Gao*)

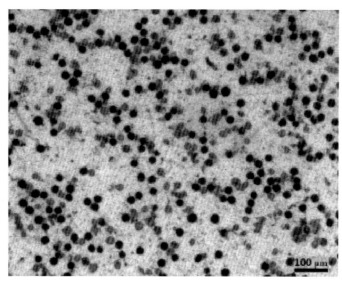

图2 2016—2017年杂交季节CP01-1178花粉镜检结果

Fig. 2 Microscopic pollen examination of CP01-1178 in 2017—2018 crossing season.

广西斑茅表型性状遗传多样性分析

为有效评价和利用广西斑茅种质资源，扩增甘蔗遗传基础。对183份广西斑茅种质资源主要表型性状及遗传多样性进行分析（图3）。结果表明：广西斑茅种质资源表型遗传多样性比较低，13个描述型性状的遗传多样性指数在0.000 0～1.234 9，平均为0.307 0，以毛群较高，生长带形状较低，空心、气根、根点排列和脱叶性4个性状无多态性表现；不同地区的斑茅资源遗传多样性指数在0.285 1～0.507 2，且以钦州的多样性最大，其次是桂林和崇左，以来宾的多样性最小。5个数值型性状的变异系数在13.54%～29.11%，平均为19.59%，以叶宽比较大，叶长较小；10个地区的斑茅资源变异系数在16.48%～21.92%，以桂林最大，百色最小。通过聚类分析，183份资源可以分为10个类群，各类群遗传分化不明显，与地理来源无密切联系。本研究揭示了广西不同地区斑茅的表型特异性和遗传多样性，为斑茅资源的采集、保育和杂交利用提供理论参考。

(黄玉新，张保青，高轶静，段维兴*，周珊，杨翠芳，张革民，王泽平)

Phenotypic traits and genetic diversity of *Erianthus arundinaceum* germplasm from Guangxi

This study was conducted to effectively evaluate and utilize the germplasm resources of *Erianthus arundinaceum* in Guangxi, to expand the genetic basis of sugarcane. The phenotypic traits and genetic diversity of 183 accessions of *E. arundinaceum* collected from Guangxi were analyzed (Fig.3). Results showed that the genetic diversity of phenotypic traits was low. The genetic diversity index (DI) ranged of 13 descriptive traits from 0.000 0 to 1.234 9 with an average of 0.307 0, with highest for hair groups and lowest for growth ring shape. There was no polymorphism in 4 traits including pipe, aerial root, root primordial and sheath detached. The DI among different regions ranged from 0.285 1 to 0.507 2, with highest for Qinzhou, followed by Guilin and Chongzuo, and the lowest for Laibin. The coefficient of variation (CV) of 5 numerical traits ranged from 13.54% to 29.11% with an average of 19.59%, with highest for leaf width and lowest for leaf length. The regional CV of 10 regions ranged from 16.48% to 21.92%, with highest for Guilin and the lowest for Baise. Cluster analysis showed that 183 germplasm resources could be clustered into 10 groups. The genetic differentiation among groups was not obvious and had no close relationship with geographical origin and the genetic distance was not closely related to geographical origin. This study revealed the phenotypic specificity and genetic diversity of *E. arundinacea* in different regions of Guangxi, and provided theoretical reference for collection, conservation and hybridization of *E. arundinacea* resources.

（Yu-xin Huang, Bao-qing Zhang, Yi-jing Gao, Wei-xing Duan*, Shan Zhou, Cui-fang Yang, Ge-min Zhang, Ze-ping Wang）

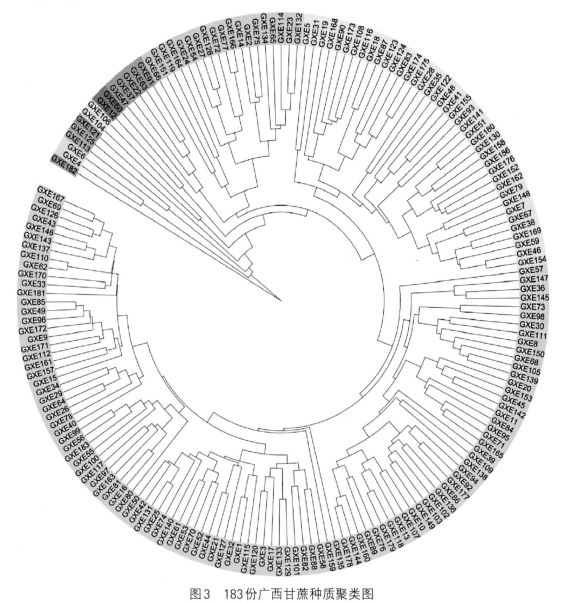

图3　183份广西甘蔗种质聚类图

Fig. 3　Cluster analysis for 183 accessions of *E. arundiraceum* from Guangxi

138份国外引进甘蔗品种（系）宿根性评价

了解国外引进甘蔗种质的宿根蔗特性，为筛选宿根性强、高产高糖、综合性状优的亲本提供依据（表10）。以ROC22为对照，对138份引进品种（系）（美国79份、法国49份、菲律宾10份）第二年宿根株高、茎径、单茎重、锤度、有效茎数和黑穗病自然发病情况进行评价分析。3个国家品种（系）宿根蔗综合农艺性状优劣比较为美国＞法国＞菲律宾；有30个品种（系）感染黑穗病，占参试材料的21.7%，平均感病率大小依次为法国＞美国＞菲律宾；根据农艺性状聚类分析结果表明：包括ROC22在内的139份材料可以分为3个大类群，6个亚类群及7个小类群；第Ⅲ类群的A2和B亚群整体表现优于其他类群品种（系）。美国品种（系）CP89-176、CP51-21、CP88-1834和CP67-412、法国品种（系）FR93-774和FR97-164、菲律宾品种（系）VMC71-39等宿根农艺性状表现突出，可考虑作为高产高糖、强宿根亲本。

（黄玉新，段维兴*，张保青，杨翠芳，高铁静，周珊，张革民，李翔）

Evaluation of ratoon characteristics of 138 exotic sugarcane germplasm

To provide reference basis for parent selection with good ratoon performance, high yield and high sugar content, the ratoon characters of sugarcane germplasm introduced from abroad were evaluated (Table 10). Six indicesindices of the second ratoon including stalk length, stalk diameter, weight per stem, brix, millable stalks and smut disease incidence for 138 introduced sugarcane germplasm (79 from U.S.A., 49 from France and 10 from Philippines) were analyzed with ROC22 as control. The comprehensive agronomic character analysis of the ratoon cane showed the propensity of ratooning in the following order; germplasm from U.S.A.>France> Philippines. Thirty varieties, about 21.7% of the test clones, were infected with smut at varying degrees. The smut disease incidence rating was France> U.S.A.>Philippines. Cluster analysis showed that 139 clones evaluated could be divided into 3 groups, 6 sub-groups and 7 minor groups. The A2 small groups and B subgroups within class III showed better comprehensive performance than other groups. CP89-176, CP51-21, CP88-1834, CP67-412, FR93-774, FR97-164 and VMC71-39 had best ratoon performance for cane yield, sucrose content and ratoonability, making them as desirable breeding parents.

(Yu-xin Huang, Wei-xing Duan*, Bao-qing Zhang, Cui-fang Yang, Yi-jing Gao, Shan Zhou, Ge-min Zhang, Xiang Li)

表10 优异材料综合性状表现

Table 10 The performance of agronomic traits in superior clones

供试材料 Name	来源 Origin	株高（cm） Stalk Length	茎径（cm） Stalk Diameter	单茎重（kg） Weight per stalk	锤度 (°Brix)	有效茎数（条） Millable stalks	黑穗病发病率（%） Incidence of smut
ROC22 (CK)		312.4	2.72	1.815	23.36	32	6.3
CP33-485		369.0	2.29	1.517	12.74	58	0.0
CP89-2376		340.0	2.80	2.100	18.06	31	0.0
CP89-170	美国 America	276.4	2.93	1.866	20.30	68	0.0
CP6		343.4	2.82	2.139	23.12	26	0.0
CP94-1340		343.0	2.94	2.322	22.26	32	3.1
CP89-176		278.2	3.08	2.067	21.74	49	0.0

(续)

供试材料 Name	来源 Origin	株高（cm） Stalk Length	茎径（cm） Stalk Diameter	单茎重（kg） Weight per stalk	锤度 (°Brix)	有效茎数（条） Millable stalks	黑穗病发病率（%） Incidence of smut
Hocp01-564	美国 America	305.0	1.95	0.915	23.18	66	0.0
Hocp92-64		293.0	2.94	1.984	20.04	34	0.0
FR93-774	法国 France	314.0	2.90	2.080	23.16	50	0.0
FR99-413		243.8	2.64	1.337	24.64	28	0.0
FR93-816		360.0	2.33	1.540	20.76	47	4.3
FR93-257		280.0	2.99	1.961	22.14	27	0.0
FR96-418		301.0	3.24	2.488	21.34	29	0.0
FR01-03		264.0	2.95	1.800	23.48	18	0.0
FR97-164		301.0	2.17	1.109	23.68	55	0.0
FR97-80		367.0	2.95	2.505	22.70	34	8.8
FR96-22		351.4	2.87	2.270	21.40	41	0.0
VMC88-354	菲律宾 Philippines	394.0	2.11	1.380	15.58	94	0.0
VMC95-88		352.6	3.36	3.134	20.82	24	0.0
VMC95-37		314.4	2.85	2.006	16.44	30	0.0
VMC96-60		369.0	2.86	2.364	17.48	44	0.0
VMC73-229		356.0	2.88	2.316	20.40	18	11.1

斑茅割手密复合体杂交利用过程野生特异基因遗传分析

揭示斑茅割手密复合体在杂交利用过程中的斑茅、割手密野生特异基因在各世代的遗传规律，为利用斑割复合体创制甘蔗育种新亲本提供理论依据（图4）。利用AFLP-PCR分子标记结合毛细管电泳技术对斑割复合体在杂交利用过程中的斑茅、割手密野生特异基因在各世代的传递动态进行分析，并研究它们之间的遗传关系。29对AFLP引物组合共扩增出3 695个位点，多态性比例为97.89%。斑茅和割手密对斑割复合体的遗传贡献率分别为43.96%和56.04%。斑茅特异位点在F_1、BC_1和BC_2 3个世代的平均遗传率分别为8.25%、1.90%和0.63%，割手密特异位点在F_1、BC_1和BC_2 3个世代的平均遗传率分别为16.98%、2.40%和0.21%，特异遗传物质均呈逐代减少趋势。比较不同世代甘蔗栽培种亲本遗传到后代的特异位点比率，F_1的GT02-761特异位点比率最高，BC_1的GT05-2743特异位点比例平均高达92.75%，BC_2的ROC_2 3遗传率最低，为49.09%，FN39遗传率最高，达94.32%。聚类分析结果表明斑割复合体偏向父本遗传，斑割复合体杂交后代偏向甘蔗栽培种遗传，与分子遗传关系分析结果一致。研究表明，经过3代的遗传重组，斑割复合体后代的遗传物质与斑割复合体相比已发生了很大的改变；研究明确了斑茅、割手密2个亲本在3个世代的遗传贡献规律，为进一步的杂交选育提供理论支持。

（周珊，高轶静，张保青，黄玉新，段维兴，杨翠芳，王泽平，张革民*）

Genetic analysis of specific genetic loci in *Erianthus arundinaceus*×*Saccharum spontaneum* introgression clones.

The inheritance of specific genetic loci of *Erianthus arundinaceus* (Retz.) Jeswiet and *Saccharum*

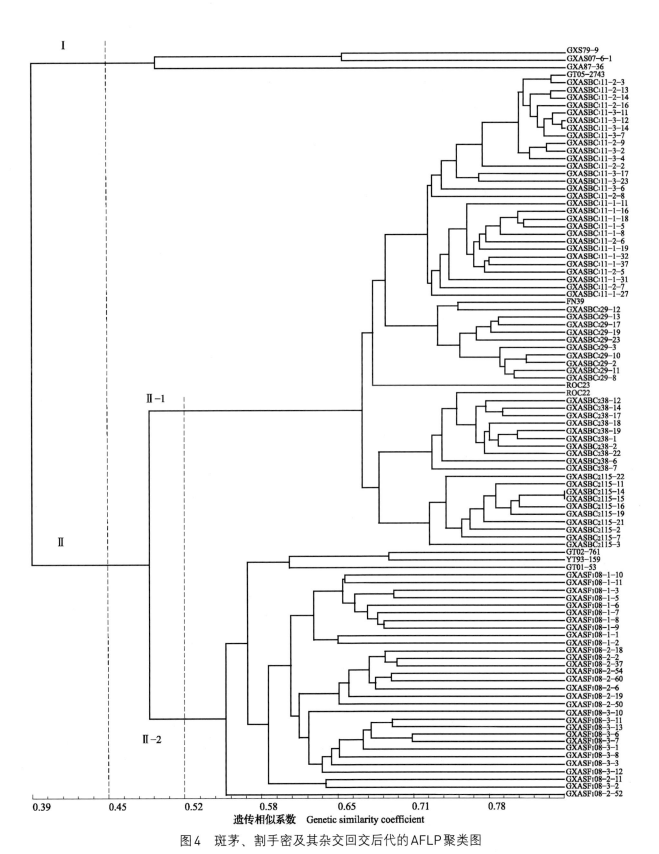

图 4 斑茅、割手密及其杂交回交后代的 AFLP 聚类图

Fig. 4 Genetic similarity of GXA87-36, GXS79-9 and their derivatives determined using AFLP markers

spontaneum L. in intergeneric hybrids of *E. arundinaceus*×*S. spontaneum* was investigate, in order to provide theoretical basis for producing and using new sugarcane parents (Fig. 4). AFLP-PCR molecular markers combined with capillary electrophoresis were used to analyze the inheritance of specific loci over multiple generations and established the genetic relationship among these materials. A total of 3695 loci were amplified from 29 AFLP primer combinations, with a polymorphic ratio of 97.89%. The genetic fragments derived from *E. arundinaceus* (Retz.) Jeswiet and *S. spontaneum* L. were 43.96% and 56.04%, respectively. The average heritability of *E. arundinaceus* (Retz.) Jeswiet in F_1, BC_1 and BC_2 were 8.25%, 1.9% and 0.63%, respectively, and the average heritability of *S. spontaneum* L.in F_1, BC_1 and BC_2 was 16.98%, 2.4% and 0.21% respectively. This indicates a reduction of introgressed genetic loci in successive generations. The analysis of sugarcane specific loci in different generations showed that, the genetic complement of GT02-761 was retained most in F_1. The proportion of specific loci from GT05-2743 inheriting to BC_1 progeny was 92.75%. The heritability for ROC23 inheriting to BC_2 was the lowest, and the heritability for FN39 was the highest, up to 94.32%. The cluster analysis showed that the intergeneric hybrid complex inherited more genetic information from male parent and their progeny inherited more genetic material from sugarcane (*Saccharum* spp.). This was consistent with the analysis of molecular genetic relationship. The results showed that the genetic loci of hybrids from intergeneric hybrids had changed greatly compared with of the progeny of intergeneric hybrids after three generations of back-crossing. Thus, this work analyzed the inheritance of genetic materials from introgression parents for up to three generations, and this knowledge provide direction for future genetic base-broadening and breeding.

(Shan Zhou, Yi-jing Gao, Bao-qing Zhang, Yu-xin Huang, Wei-xing Duan, Cui-fang Yang, Ze-ping Wang, Ge-min Zhang[*])

河八王杂交种F_1、BC_1及其亲本DNA甲基化水平和模式变化

为分析河八王及后代基因组DNA甲基化水平和遗传模式变化，以河八王及其后代F_1和BC_1为材料（表11），采用甲基化敏感扩增多态性技术（Methylation sensitive amplification polymorphism, MSAP）结合毛细管电泳技术（Capillary electrophoresis, CE）分析亲本及后代基因组DNA甲基化水平和遗传模式变化规律。结果表明：母本'GT05-3256'的MSAP比率是59.6%，父本'GXN1'则是60.5%，其杂交种F_1的MSAP比率是56.4%～59.0%，均低于双亲；BC_1的母本'YC94-46'的MSAP比率是59.4%，父本'T6-3'则是59.0%，BC_1 MSAP比率是56.9%～69.8%，整体平均为62.8%，平均值高于双亲。BC_1世代总甲基化水平略高于F_1世代水平。F_1和BC_1基因组CCGG位点发生甲基化的方式整体上以内部胞嘧啶双链甲基化为主。在F_1和BC_1中均检测到70种甲基化类型，并进一步分为A、B、C、D和E等5大类，其中A类是亲本向杂交种的甲基化遗传类型；B类是去甲基化类型，表示杂交种相应于其亲本的甲基化减弱；C类是过或超甲基化类型，表示杂交种相应于其亲本的甲基化增强；D类是次甲基化类型，杂交后代甲基化水平比双亲均要低；E类为不定类型。结果显示，B、C、D、E是杂交种的甲基化变异类型；F_1的甲基化遗传类型（A类）比例明显低于BC_1，但变异类型B、C、D、E高于BC_1；杂交形成F_1和BC_1过程中，基因组DNA普遍发生了超甲基化

修饰。

（刘昔辉，桂意云，张荣华，李海碧，韦金菊，周会，杨荣仲，张小秋，李杨瑞*）

DNA methylation levels and genetic patterns in F_1 and BC_1 progeny of *Narenga porphyrocoma*（Hance）and the clones used for crossing

DNA methylation levels and genetic patterns in F_1, BC_1 and their parents were detected using methylation sensitive amplification polymorphism (MSAP) and capillary electrophoresis technology (CE) (Table 11). The results showed that MSAP ratio in the maternal parent GT05-3256 was 59.6%, 60.5% in the paternal parent GXN1, and 56.4% ~ 59.0% in in the hybrid F_1, showing lower than that in both parents. The ratio of MSAP was 59.4% in the maternal parent YC94-46, and 59.0% in the paternal parent T6-3, 56.9% ~ 69.8% in BC_1 with the average of 62.8%, showing higher than that in both parents. The total methylation level in the BC_1 generation is slightly higher than that in the F_1 generation. The methylation of CCGG loci in sugarcane genome of the individuals in F_1 and BC_1 mainly occurred on internal cytosine double-strand. Seventy-one different methylations patterns were detected in the F_1 and BC_1 populations, and they were further divided into 5 types: type A, the methylation patterns in the hybrids were the same as their parents; type B, demethylation, the methylation of many genetic loci in the hybrids was lost compared to that of the parents; type C, hypermethylation, the methylation in the hybrids was enhanced corresponding to the parents; type D, hypomethylation, the degree of methylation in the hybrids was lower than that in the both parents; and type E, uncertain type. Among of them, type A is the genetic transfer type from the parent to the hybrid, and types B, C, D and E are the methylation variants of the hybrids. The results showed that the rate of type A methylation in the F_1 individuals was significantly lower than that in the BC_1, but B, C, D, and E types were higher than that in the BC_1. During the F_1 and BC_1 hybridization, the hypermethylation occurred in the whole genome.

(Xi-hui Liu, Yi-yun Gui, Rong-hua Zhang, Hai-bi Li, Jin-ju Wei, Hui Zhou, Rong-zhong Yang, Xiao-qiu Zhang, Yang-rui Li*)

表11 杂交种F_1、BC_1及亲本的甲基化水平比较

Table 11 Methylation in F_1, BC_1 and parents

材料 Material	酶切条带类型 Types of digestion band				总扩增位点 Total amplified loci	总扩增带数 Total amplified bends	总甲基化带数 Total methylated bends	甲基化敏感扩增多态性（%）MSAP	全甲基化率（%）Total methylation rate	半甲基化率（%）Half methylation rate
	I	II	III	IV						
3256	260	220	163	778	1 421	643	383	59.6	34.2	25.3
GXN1	236	190	172	823	1 421	598	362	60.5	31.8	28.8
T6-1	282	203	166	770	1 421	651	369	56.7	31.2	25.5
T6-4	278	215	145	783	1 421	638	360	56.4	33.7	22.7
T6-5	237	172	155	857	1 421	564	327	58	30.5	27.5
T6-3	280	253	150	738	1 421	683	403	59	37	22

(续)

材料 Material	酶切条带类型 Types of digestion band				总扩增位点 Total amplified loci	总扩增带数 Total amplified bends	总甲基化带数 Total methylated bends	甲基化敏感扩增多态性（%） MSAP	全甲基化率（%） Total methylation rate	半甲基化率（%） Half methylation rate
	I	II	III	IV						
94-46	259	213	166	783	1 421	638	379	59.4	33.4	26
H4-4	273	160	201	787	1 421	634	361	56.9	25.2	31.7
H4-11	252	219	170	780	1 421	641	389	60.7	34.2	26.5
H4-13	226	222	150	823	1 421	598	372	62.2	37.1	25.1
H4-27	225	159	164	873	1 421	548	323	58.9	29	29.9
H4-29	222	164	229	806	1 421	615	393	63.9	26.7	37.2
H4-43	187	188	244	802	1 421	619	432	69.8	30.4	39.4
H4-47	196	264	142	819	1 421	602	406	67.4	43.9	23.6

5.1.2 甘蔗高效育种 Efficient Sugarcane Breeding

桂糖甘蔗新品系黑穗病抗性鉴定及结果分析

防治甘蔗黑穗病最有效的途径是种植抗病品种，而甘蔗抗黑穗病评价则是抗病品种选育过程中重要程序（表12）。本研究对广西农科院甘蔗研究所选育的8个桂糖甘蔗新品系进行人工浸渍接种甘蔗鞭黑粉菌混合冬孢子悬浮液，同时调查大田自然发病情况，收集一新一宿黑穗病发病情况，以新台糖22号为对照品种，综合评价桂糖甘蔗新品系对黑穗病的抗性。综合新植、宿根的人工接种和大田自然发病评价结果，抗性类型为抗病的品系有2个，分别为桂糖12-765和桂糖12-2262；抗性类型为中抗的品系有2个，分别为桂糖12-2476和桂糖12-2004；抗性类型为中感的品系有1个，为桂糖12-162；抗性类型为感病的品系有3个，分别为桂糖12-762、桂糖12-2425和桂糖12-917；对照品种新台糖22号的综合抗性类型为感病。对比人工接种和自然发病结果，人工接种黑穗病能更准确评价甘蔗品种抗黑穗病的水平，为选育高产高糖高抗黑穗病甘蔗品种提供依据。

（经艳，周会，刘昔辉，谭芳，张小秋，张荣华，宋修鹏，李杨瑞，颜梅新，雷敬超，李海碧，覃振强，罗亚伟，李冬梅，韦金菊[*]）

Analysis of results from smut resistant identification in new sugarcane clones of Guitang

The most effective way to control sugarcane smut is to plant resistant varieties, and the evaluation of resistance to sugarcane smut is a very important procedure in the breeding of resistant varieties (Table 12). In order to evaluate the resistance to smut of the eight new sugarcane clones bred by Sugarcane Research Institute of Guangxi Academy of Agricultural Sciences, the new clones were inoculated with a suspension of Sporisorium scitamineum to screen for smut incidence. Also, the natural incidences of smut in the field were investigated. The Investigation period included new planting and first ratoon of sugarcane. ROC22 was used as control. The results showed that GT12-765 and GT12-2262 was

表12 桂糖甘蔗新品系黑穗病抗性鉴定结果

Table 12 Identification results of smut resistance of new GT clones

品系 Clones	亲本组合 Parents	处理 Treatment	新植 New plant					宿根 Ratoon			新宿对比 Comparison of new plant and ratoon	综合抗性水平 Comprehensive resistance level
			最高发病率（%）Maximum incidence	抗性级别 Resistance grade	抗性类型 Resistance types	潜伏期（d）Incubation period	最高发病率（%）Maximum incidence	抗性级别 Resistance grade	抗性类型 Resistance types			
桂糖12-765 GT12-765	RB83-5089×桂糖02-901	人工接种	8.59	3	R	63	7.03	2	R	+1	R
		自然发病	0	1	HR	—	3.81	1	HR	0	
桂糖12-762 GT12-762	桂糖97-182×粤糖91-976	人工接种	20.17	5	MS	46	36.76	6	S	−1	S
		自然发病	4.17	2	R	63	13.26	3	R	−1	
桂糖12-2425 GT12-2425	桂糖94-119×ROC23	人工接种	18.97	5	MS	63	44.98	7	S	−2	S
		自然发病	6.32	2	R	63	16.81	4	MR	−2	
桂糖12-2476 GT12-2476	粤糖85-177×桂糖73-167	人工接种	12.07	4	MR	63	17.24	4	MR	0	MR
		自然发病	0.85	1	HR	115	2.47	1	HR	0	
桂糖12-2004 GT12-2004	科5×桂糖02-901	人工接种	9.68	4	MR	78	17.39	4	MR	0	MR
		自然发病	0	1	HR	—	0.90	1	HR	0	
桂糖12-917 GT12-917	桂糖02-208×粤糖00-236	人工接种	26.58	6	S	46	22.57	5	MS	+1	S
		自然发病	0	1	HR	—	4.46	1	HR	0	
桂糖12-2262 GT12-2262	桂糖04-107×桂糖04-120	人工接种	2.91	2	HR	115	6.60	2	R	0	R
		自然发病	0	1	HR	—	8.53	2	R	−1	
桂糖12-162 GT12-162	粤糖94-128×农林8	人工接种	13.25	5	MS	46	8.05	2	R	+3	MS
		自然发病	0	1	HR	—	4.88	1	HR	0	
ROC22	ROC5号×69-463	人工接种	13.00	5	MS	46	37.36	6	S	−1	S
		自然发病	6.35	2	R	63	41.54	7	S	−5	

注："+"表示宿根抗病性水平提高；"−"表示宿根抗病性水平降低；"0"表示新宿根抗性水平不变；"+"或"−"后面的数字表示抗性变化的幅度。

Note: "+" indicates that the resistance level of ratoon was advanced; "−" indicates that the resistance level of ratoon was lower; "0" indicates that the resistance level of new plant and ratoon were same; The number of behind in "+" or "−" mean was the range of resistance level change.

resistant to smut; the clones GT12-2476 and GT12-2004 were moderately resistant; GT12-162 was moderately susceptible; GT12-762, GT12-2425 and GT12-917 were susceptible to smut; and the CK (ROC22) was susceptible. The artificial inoculation could evaluate smut resistance more accurately , compared to the natural infection. The results of this study provided a basis for breeding for smut resistant sugarcane varieties.

(Yan Jing, Hui Zhou, Xi-hui Liu, Fang Tan, Xiao-qiu Zhang, Rong-hua Zhang, Xiu-peng Song, Yang-rui Li, Mei-xin Yan, Jing-chao Lei, Hai-bi Li, Zhen-qiang Qin, Ya-wei Luo, Dong-mei Li, Jin-ju Wei[*])

5.2 甘蔗栽培及生理
Sugarcane Cultivation and Physiology

5.2.1 甘蔗生理生态研究
Study on Physiology and Ecology of Sugarcane

The impact of silicon on photosynthetic and biochemical responses of sugarcane under different soil moisture levels

Increasing drought stress is one of the most limiting factors for agricultural crop productivity across the world. Silicon (Si) has been known to augment plant protection against drought stress (Fig.5). In this experiment, the responses of sugarcane with silicon application to drought stress for photosynthetic and biochemical activities were investigated. Three water regimes [75 ± 5, 50 ± 5and 25 ± 5% of soil water content capacity (SWCC) from 70 to 115 days after transplanting] and six silicon levels such as 0, 20, 40, 60, 80 and 100 g $CaO.SiO_2$ pot^{-1} equivalent to 0, 194, 387, 581, 774 and 968 mg Si kg^{-1} soil, respectively, were applied. This experiment was arranged in a completely randomized block design. It was found that water stressed plants with silicon application showed increasing trends in photosynthetic CO_2 assimilation (~2% –106%), stomatal conductance (~8% –113%), transpiration (~13% –274%) and chlorophyll fluorescence yield (~0.3% –10%) of 75 ± 5, 50 ± 5 and 25 ± 5% of SWCC as compared to the controls without Si application. The silicon application improved the plant growth under water stress, accompanied with the up-regulation in leaf relative water content (~2%–8%), photosynthetic pigments (~2%–35%), activities of catalase (ca.12%–91%), peroxidase (ca.7%–30%) and superoxide dismutase (ca.3% –96%) enzymes. These results indicate that Si fertilizer plays an important role in mitigating the negative effects of drought stress on sugarcane plant growth by improving water status, photosynthetic parameters and activating the antioxidant machinery. Findings demonstrated the silicon application an efficient strategy to improve sugarcane tolerance to drought stress.

(Krishan K. Verma, Xi-hui Liu, Kai-chao Wu, Rajesh Kumar Singh, Qi-qi Song, Mukesh Kumar Malviya, Xiu-peng Song, Pratiksha Singh, Chhedi Lal Verma, Yang-rui Li[*])

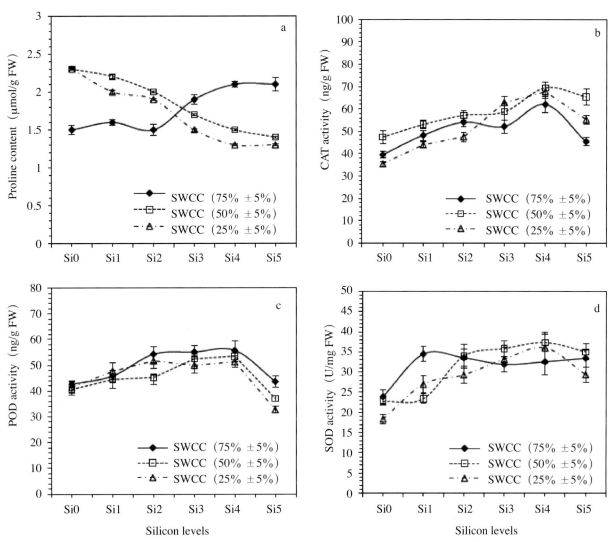

Fig. 5 Ameliorating effect of silicon on the content of proline (a), and the activities of catalase (b), peroxidase (c) and superoxide dismutase (d) in sugarcane leaves grownin different soil water content capacity.

Impact of seed coating agents on single-bud seedcane germination and plant growth in commercial sugarcane cultivation.

China is the third largest sugar-producing country in the world. The cost of sugarcane planting has increased rapidly in recent years, and the existing planting model needs to be changed to reduce the cost (Table 13). The aim of this study was to evaluate the effects of commercial seed coating agents on the germination and yield of sugarcane. Six commercial seed coating agents (Premis, Gaucho, Colest, Dividend, Manshijin and Maishuping) used for other crops were used to coat the 4-cm-long single seedcane setts and stored for 7 days, respectively, at 22°C before planting. It was found that all the seed coating agents were effective to reduce the water evaporation of seedcane setts and ensured better germination rate, and Premis, Gaucho, Colest and Maishuping were found better than Dividend and Manshijin. The economic traits of sugarcane, plant height, stalk diameter, single stalk weight and Brix were significantly higher in the Premis seedcane coating treatment than the control at maturity stage in a green house experiment. These results would be helpful to promote the development of the mechanical planting of single-bud seedcane setts for increasing sugarcane production.

(Yong-jian Liang, Xiao-qiu Zhang, Liu Yang, Xi-hui Liu, Li-tao Yang[*], Yang-rui Li[*])

Table 13 Seed coatings used for seedcane setts screening

Product name and nature of seed coating agents	Manufacturer	Active ingredient content (g/L)
Premis (Triticonazole)	BASF, Germany	25
Gaucho (Imidacloprid)	Bayer, Germany	600
Colest (Fludioxonil)	Syngenta, Switzerland	25
Dividend (Difenoconazole)	Syngenta, Switzerland	30
Manshijin (Fludioxonil, Mefenoxam)	Syngenta, Switzerland	35
Maishuping (Thiamethoxam, Fludioxonil, Mefenoxam)	Syngenta, Switzerland	25
Control (water)		

Physiological and molecular analysis of sugarcane (varieties-F134 and NCo310) during *Sporisorium scitamineum* interaction

Sugarcane smut is a prevalent fungal disease, which causes considerable cane yield loss in China. However, limited information is available for sugarcane infected by the pathogen Sporisorium scitamineum at the different developmental stages (Fig.6). The present study evaluated the physiological, cytological, biochemical, and molecular changes in two sugarcane varieties: F134 and NCo310, at consecutive time intervals (30–180 days) after S. scitamineum infection. qRT-PCR examination results showed that the expression level of chitinase (*ScChi*) was higher (4, 5, 6, and 40 times) in F134 and that of glucanase (*ScGluD*1) was 2 times higher at 30, 90 and 180 days as compared to NCo310. Similarly, the elevated levels of the hormone GA3, ABA, and IAA showed incompatible results in leaf and root tissues in both sugarcane varieties. The results confirmed the disease resistance in sugarcane variety F134 against pathogen S. scitamineum after the interaction. In addition, SDS-PAGE assay showed the induction of some protein bands in infected NCo310 at 60 days, and anatomical changes in the nucleus and chloroplast structure in stem tissues were also confirmed by TEM, showing highly disrupted in NCo310 as compared to F134. All physiological parameters decreased more in NCo310 than F134, as compared to control. The overall studies revealed that the variety NCo310 had higher disease severity than F134. The present findings will shed new light on the role of differential expression of stress-related genes and biochemical changes playing important role in sugarcane smut infection.

(Pratiksha Singh, Qi-qi Song, Rajesh Kumar Singh, Hai-bi Li, Manoj Kumar Solanki, Li-tao Yang[*], Yang-rui Li[*])

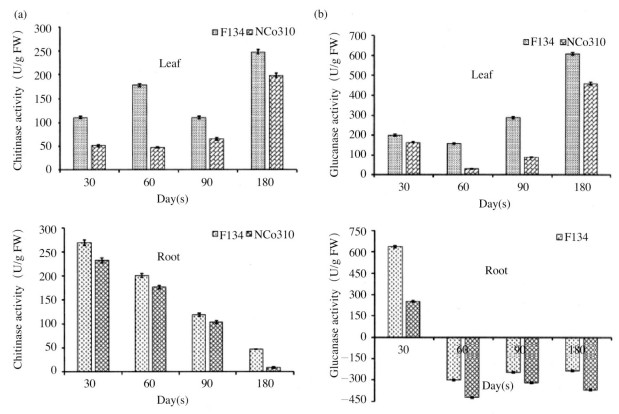

Fig. 6 Changes in enzyme activities in leaf and root tissues of sugarcane varieties (F134 and NCo310) at different time intervals after *S.scitamineum* inoculation. a Chitinase; b glucanase. All data points (with the deduction of their controls) are the mean ± SE (n = 3)

Evaluation of lodging resistance in sugarcane (*Saccharum* spp. hybrid) germplasm resources

In order to determine the lodging propensity of different sugarcane varities, a field experiment was conducted in 2018. Thirty sugarcane clones were used as the the lodging grade, fracture resistance force, basal stem diameter, middle stem diameter and brix were investigated (Table 14). With the data of lodging classification, the lodging resistance index was established. It was found that the lodging resistance index was significantly and positively correlated with the ratio of basal and middle stem diameters and brix (Fig. 7). A cluster analysis was conducted, for the thirty sugarcane combinations by the lodging resistance index, which showed sixteen were lodging resistant, thirteen half-lodging, and two complete-lodging.

(Xiang Li, Yi-jie Li, Qiang Liang, Shang-Hai Lin, Qu-yan Huang, Rong-zhong Yang, Li-tao Yang[*], Yang-rui Li[*])

Table 14 Correlation coefficients between different traits

Correlation coefficient	Lodging resistance index	Resistance force	The ratio of middle in basal stem diameter	°Brix	Basal stem diameter	Middle stem diameter
Lodging resistance index	1					
Resistance force	−0.02	1				

（续）

Correlation coefficient	Lodging resistance index	Resistance force	The ratio of middle in basal stem diameter	°Brix	Basal stem diameter	Middle stem diameter
The ratio of middle in basal stem diameter	0.201*	-0.037	1			
°Brix	0.232*	0.052	-0.168	1		
Basal stem diameter	0.419**	0.294**	0.432	0.093	1	
Middle stem diameter	0.356**	0.223*	-0.049	0.19	0.87	1

*and** represent the significance at 5%, and 1% levels, respectively

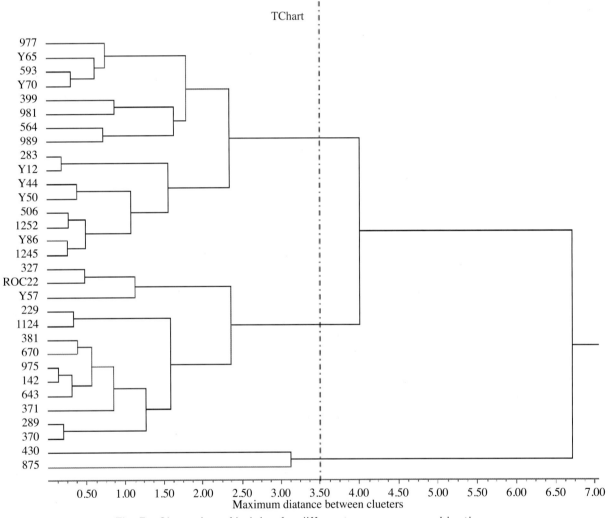

Fig. 7　Clustering of lodging for different sugarcane combinations

不同施氮水平下甘蔗内源激素、产量和糖分的变化特征

桂糖42号是我国甘蔗推广面积最大的自育品种，探讨该品种的氮肥需求特性，可以为指导该品种在蔗区科学管理提供理论依据（表15、表16）。以桂糖42号为供试材料，在分蘖前施用不同水平的氮肥（CK：尿素0 kg/hm²；低水平：尿素150 kg/hm²，300 kg/hm²；中水平：尿素450

kg/hm²；高水平：尿素600 kg/hm²，750 kg/hm²），研究不同氮肥水平甘蔗内源激素含量（GA、CTK、BR 和 ABA）、产量和糖分的变化特征。结果表明，不同施肥水平下甘蔗产量顺序为150 kg/hm² > 300kg/hm² > 0kg/hm² > 450 kg/hm² > 600 kg/hm² > 750 kg/hm²，其中150 kg/hm²和300 kg/hm²分别比对照提高了5.40%和2.54%（$P<0.01$），超过300 kg/hm²施肥量后甘蔗产量呈下降趋势；不同施肥量下的每公顷含糖量从大到小顺序与产量顺序相同但施氮肥多少不同不会对糖分产生影响；经济效益顺序为150 kg/hm² > 0 kg/hm² > 300 kg/hm² > 450 kg/hm² > 600 kg/hm² >750 kg/hm²；不同施氮肥水平下GA含量和BR含量的表现一致，在150 kg/hm²施肥条件下达到最高值，随后施氮肥量的增加而呈下降趋势；ABA则呈相反趋势，即含量随着施肥量的增加而增加；而CTK则呈先升高后下降再升高的趋势。

（周慧文，陈荣发，范业赓，丘立杭，李杨瑞，吴建明*，黄杏*）

The effect of different nitrogen fertilization rates on endogenous hormone and quality content of sugarcane

The sugarcane variety GT42 is a locally-bred variety, and it is the most cultivated clone in China. It is necessary to explore the nitrogen demand of the variety GT42 and it could provide reference for scientific management of the variety in sugarcane planting area (Table 15 and Table16). The sugarcane variety GT42 was grown with different levels of nitrogenous fertilizer (CK: 0 kg/hm²; low level: carbamide (urea) 150 kg/hm², 300 kg/hm²; middle level: carbamide 450 kg/hm²; high level: carbamide: 600 kg/hm², 750 kg/hm²) before the tiller stage, and the content of endogenous hormones, final yield and sugar content were measured n the tillering stage. The results show that the yield under different levels nitrogenous fertilizer in decreasing order was: 150 kg/hm²>300 kg/hm²>0 kg/hm²>450 kg/hm²>600 kg/hm²>750 kg/hm². The yield under 150 kg/hm² and 300 kg/hm² fertilizer levels were 5.40% and 2.54% ($P<0.01$) more than the CK, respectively. And the yield of GT42 reached the maximum with 150 kg/hm² of nitrogen fertilization. The sugar content per hectare followed the same order as the yield but the nitrogen fertilization levels had no significant effect on the sucrose content. The order of economic benefit from higher to lower was: 150 kg/hm²>0kg/hm²>300 kg/hm²450 kg/hm²>600 kg/hm²>750 kg/hm². The variation of GA and BR content was the same that reached the maximum with 150 kg/hm² of nitrogen fertilization level and then decreased with increasingnitrogen fertilization level. The content of ABA showed the opposite trend, increasing with the increase of nitrogen fertilization level. The content of CTK was increased first and then decreased and then increased.

(Hui-wen Zhou, Rong-fa Chen, Ye-geng Fan, Li-hang Qiu, Yang-rui Li, Jian-ming Wu*, Xing Huang*)

表15　不同施氮肥水平下的甘蔗品质

Table15　The quality of sugarcane in different nitrogen fertilization levels

项目 Iterm	施氮肥水平 (kg/hm²) Nitrogen fertilization levels					
	0	150	300	450	600	750
产量(kg/hm²) Yield	83 980.5ab	88 516.5a	86 054.85a	81 171.45ab	80 992.8b	80 317.95b
糖分（%）Sugarcrose content	14.49a	13.99a	14.15a	14.97 a	14.46a	14.49a

(续)

项目 Item	施氮肥水平 (kg/hm²) Nitrogen fertilization levels					
	0	150	300	450	600	750
纯度（%）Purity	89.70a	88.09 a	89.23a	89.43a	88.30a	87.95a
每公顷含糖量（%）Sugarcrose content per hectare	12 168.88a	12 383.46a	12 176.64a	12 151.37a	11 711.56b	11 638.07 b

注：同行不同小写字母表示各处理间差异显著（$P<0.05$）。
Note: Different lowercase letters in the same line indicate significant difference between treatments ($P<0.05$).

表16 不同施氮肥水平下的经济效益
Table 16 The economic benifit per hectare of the sugarcane in different nitrogen fertilization levels

项目 Iterm	施氮肥水平 (kg/hm²) Nitrogen fertilization levels					
	0	150	300	450	600	750
尿素投入 (yuan/hm²) Urea cost	0.00	815.25	1 630.50	2 445.75	33 261.00	4 076.25
施肥投入 (yuan/hm²) Fertilization cost	0.00	450.00	450.00	450.00	450.00	450.00
总投入 (yuan/hm²) Total cots	0.00	1 205.25	2 080.50	2 895.75	3 711.00	4 526.25
总收入 (yuan/hm²) Total income	41 990.25	44 258.25	43 027.50	40 585.80	40 496.40	40 159.05
净利润 (yuan/hm²) Net income	41 990.25	43 053.00	40 947.00	37 690.05	36 785.40	35 632.80

注：尿素价格为 2 500 元/t；甘蔗收入 500 元/t；机械化施肥按 450 元/hm² 计算；每公顷增加尿素投入计算公式：施肥量/含量×肥料价格/1 000；每公顷甘蔗收入＝产量×500 元/t；净利润＝甘蔗收入－总投入。
Note: The price of ureais calculated by 2 500 yuan/ton; sugarcane income is calculated by 500 yuan/ton; 450 yuan per hectare for mechanized fertilization; Calculation formula for increasing urea input per hectare=Fertilizer amount/content´fertilizer price/1 000; Sugar-cane income per hectare = yield´500 yuan/ton; Net profit = sugarcane income–total input.

甘蔗节间伸长过程赤霉素生物合成关键基因的表达及相关植物激素动态变化

以甘蔗（*Saccharum officinarum*）优良品种桂糖42号（GT42）为研究材料，分别于未伸长期（9～10叶龄以前）（Ls1）、伸长初期（12～13叶龄）（Ls2）和伸长盛期（15～16叶龄）（Ls3）取甘蔗第2片真叶（自顶部起）对应的节间组织，测定其赤霉素（GA）、生长素（IAA）、油菜素甾醇（BR）、细胞分裂素（CTK）、乙烯（ETH）和脱落酸（ABA）的含量，并通过实时荧光定量PCR（qRT-PCR）分析赤霉素合成途径关键基因 GA_{20} 氧化酶基因（*GA_{20}-Oxidase1*）、赤霉素受体基因（*GID1*）DELLA蛋白编码基因（*GAI*）的差异表达（图8、图9）。结果表明，在甘蔗伸长期间，GA和IAA含量呈现上升趋势，CTK和ABA含量呈下降趋势，ETH含量先上升后下降，BR含量则变化不明显；*GA_{20}-Oxidase1* 和GID1的表达呈上升趋势，而*GAI*的表达则呈下降趋势，这与相关植物激素的变化基本一致。综上，甘蔗节间伸长过程主要与GA和IAA相关，其次为CTK和ABA，而ETH受到IAA的调控影响节间伸长；植物激素间通过相互作用调控 *GA_{20}-Oxidase1*、*GID1* 和 *GAI* 的表达，影响GA含量和GA的信号转导过程，进而影响甘蔗节间的伸长。该研究揭示了甘蔗节间伸长过程中赤霉素生物合成途径和信号转导关键基因的差异表达及植物激素含量的动态变化规律。

（范业赓，丘立杭，黄杏，周慧文，甘崇琨，李杨瑞，杨荣仲，吴建明*，陈荣发*）

Expression analysis of key genes of gibberellin biosynthesis and related phytohormonal dynamics during sugarcane internode elongation

In this study, *sugarcane variety* GT42 was used as the research material. Second internode tissue of sugarcane at the non-elongation stage (Ls1), early elongation stage (Ls2) and elongation stage (Ls3) was collected to measure plant hormones gibberellic acid (GA), indole acetic acid (IAA), brassinosteroids (BR), cytokinin (CTK), ethylene (ETH) and abscisic acid (ABA). The differential expression of the key genes GA_{20}-*Oxidase* 1, GID1 and *GAI* was analyzed by qRT-PCR. During the elongation stage, the contents of GA and IAA showed an upward trend, but CTK and ABA contents showed a downward trend. ETH content was increased at first and then decreased, whereas BR content did not change. The expression of GA_{20}-*Oxidase* 1 and *GID* 1 increased, and that of *GAI* decreased, which was closely related to the changes in GA content (Fig. 8 and Fig. 9). Therefore, we considered that the internode elongation of sugarcane is mainly related to GA and IAA, then CTK and ABA, and ETH is regulated by IAA to affect internode elongation. The interaction between these phytohormones regulates the expression of GA_{20}-*Oxidase* 1, *GID* 1 and *GAI*, which affects the content and signal transduction of GA, thereby affecting the internode elongation of sugarcane. The study clarifies the differential expression of key genes in the gibberellin biosynthesis pathway and signal transduction process, along with the dynamic

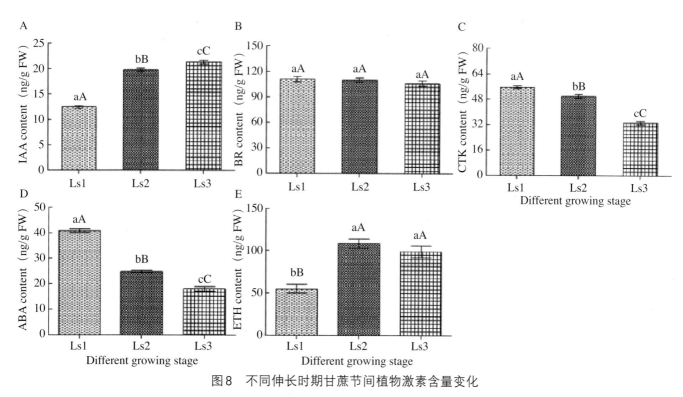

图8 不同伸长时期甘蔗节间植物激素含量变化

Fig. 8 Changes of endogenous hormone contents in sugarcane internode during different elongation stages

注：A. 生长素 (IAA), B. 油菜素甾醇 (BR), C. 细胞分裂素 (CTK), D. 脱落酸 (ABA),
E. 乙烯 (ETH); 不同小写字母表示差异显著 ($P<0.05$), 不同大写字母表示差异极显著 ($P<0.01$)。
Note: A. Indole acetic acid (IAA), B. Brassinosteroids (BR), C. Cytokinin (CTK), D. Abscisic acid (ABA),
E. Ethylene (ETH); The different lowercase letters indicate significant differences ($P<0.05$), and the different capital letters indicate extremely significant differences ($P<0.01$).

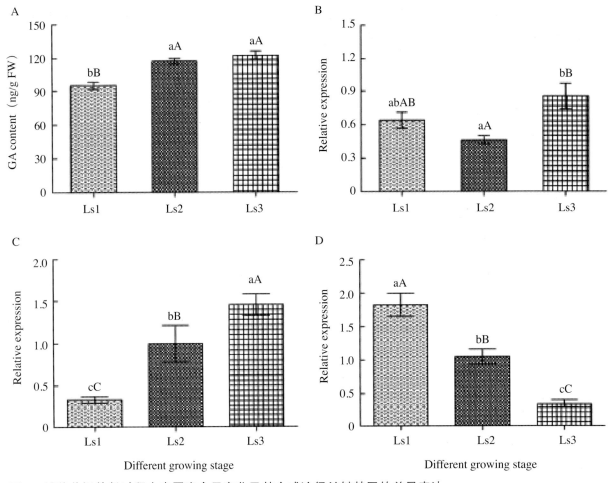

图9 甘蔗节间伸长过程中赤霉素含量变化及其合成途径关键基因的差异表达

Fig. 9　The changes of gibberellin content and genes differential expression of synthetic pathways during sugarcane inter-node elongation

注：A.GA3，B.GA_{20}-Oxidase1，C. GID 1，D. GAI；不同大写字母表示差异极显著（$P<0.01$），不同小写字母表示差异显著（$P<0.05$）。

changes of plant hormone content during internode elongation of sugarcane.

(Ye-geng Fan, Li-hang Qiu, Xing Huang, Hui-wen Zhou, Chong-kun Gan, Yang-rui Li, Rong-zhong Yang, Jian-ming Wu[*], Rong-fa Chen[*])

合理密植下强分蘖甘蔗品种性状及产量分析

甘蔗（Saccharum officinarum L.）是以收获茎秆为主的重要糖料作物，分蘖是增加有效茎进而提高甘蔗单产和产糖量的关键。本研究以强分蘖甘蔗品种"B9"为试验材料，设置4个不同种植密度（P1. 4 000芽/667m²、P2. 6 000芽/667m²、P3. 8 000芽/667m²和P4. 10 000芽/667m²），通过1年新植和1年宿根的田间小区试验，探讨种植密度对强分蘖甘蔗品种性状及产量形成的影响，并分析它们之间的相关关系（表17）。结果表明：增加种植密度，新植和宿根甘蔗分蘖盛期的总苗数呈上升趋势，但新植与宿根甘蔗总苗数的差异却在缩小，平均分别为25.55%、14.11%、8.18%和10.57%；宿根蔗株高显著大于新植蔗的株高，平均增幅分别为17.31%、10.54%、19.44%、14.88%，但茎径几乎不受种植密度的影响；新植蔗有效茎和产量随种植密度增加而增加（P2的除外），且多于宿根蔗的，但其产量却表现相反的结果，即宿根蔗总体表现增产；低密度（P1）和高密度（P4）间的锤度差异显著，并表现降低趋势。在试验条件下甘蔗产量形成过程中，种植密度与甘蔗分蘖盛期的总苗数、株高、有效茎和产量呈正相关，与茎径和田间锤度为负相关，且株高与有效茎对产量的贡献最为显著，但过多的有效茎明显不利于田间锤度提高。因此，这意味着合理密植对强分蘖甘蔗品种的产量和品质形成极其重要。

（丘立杭，范业赓，周慧文，陈荣发，黄杏，罗含敏，杨荣仲，段维兴，刘俊仙，吴建明[*]）

Analysis of agronomic traits and yield in sugarcane variety under high-density planting

Sugarcane (*Saccharum officinarum* L. inter-specific hybrids) is an important sugar. Tillering is a key trait to increase the millable canes thereby increasing yield and sugar yield per unit area. Sugarcane variety 'B9' (intense tillering ability) was planted in field plots treated with 4 different planting densities, i.e. 60 000 buds/hm^2 (P1), 90 000 buds/hm^2 (P2), 120 000 buds/hm^2 (P3), and 150 000 buds/hm^2 (P4), and the plot experiment of new planting was carried out to determine the effects of planting density on the yield characters and yield (Table 17). The results showed that, the total seedlings number of planting cane and ratoon cane in the vigorous tillering period increased with increasing planting density, but the difference between them decreased with an average of 25.55%, 14.11%, −8.18% and −10.57%, respectively. The height of ratoon cane was significantly higher than that of planting cane, with an average increase of 17.31%, 10.54%, 19.44% and 14.88%, but the stem diameter was almost un-affected by planting density. Except the result of P2, the millable canes and yield of planting cane increased with the increase of planting density, and although millable canes of planting cane was more than that of ratoon cane, but its yield showed an opposite result. The difference of brix between low planting density (P1) and high planting density (P4) was significant and showed a decreasing trend. In the process of yield formation under our test conditions, planting density was positively correlated with the total number of sugarcane seedlings at the vigorous tillering stage, plant height, millable canes and yield, negatively correlated with stem diameter and brix. The contribution of plant height and millable canes to yield was the most significant, but too many millable canes were not obviously good for brix in sugarcane. Therefore, the results indicate that the rational high-density planting is very important for the formation of maximum yield and quality of the sugarcane variety.

(Li-hang Qiu, Ye-geng Fan, Hui-wen Zhou, Rong-fa Chen, Xing Huang, Han-min Luo, Rong-zhong Yang, Wei-xing Duan, Jun-xian Liu, Jian-ming Wu[*])

表17 不同种植密度下强分蘖甘蔗成熟期有效茎数、锤度和产量

Tab.17 The millable cane, brix and yields of sugarcane variety with strong tillering under different treatment in the mature period

处理 Treatment	有效茎数 Number of millable canes in field plot		产量（kg/plot） Yield		锤度（°Brix）	
	2016年新植蔗 Plant cane in 2016	2017年宿根蔗 Ratoon cane in 2017	2016年新植蔗 Plant cane in 2016	2017年宿根蔗 Ratoon cane in 2017	2016年新植蔗 Plant cane in 2016	2017年宿根蔗 Ratoon cane in 2017
P1	258.00±8.01[bc]	247.00±20.04[c]	342.10±21.20[bc]	424.77±29.59[a]	17.33±0.74[ab]	17.89±0.41[ab]

（续）

处理 Treatment	有效茎数 Number of millable canes in field plot		产量（kg/plot） Yield		锤度（°Brix）	
	2016年新植蔗 Plant cane in 2016	2017年宿根蔗 Ratoon cane in 2017	2016年新植蔗 Plant cane in 2016	2017年宿根蔗 Ratoon cane in 2017	2016年新植蔗 Plant cane in 2016	2017年宿根蔗 Ratoon cane in 2017
P2	245.00±10.14^c	269.00±13.09^{bc}	298.37±14.47^c	391.97±42.53^{ab}	16.58±0.16^b	18.29±0.28^a
P3	291.00±4.63^{ab}	268.00±4.73^{bc}	384.73±9.88^{ab}	383.30±12.94^{ab}	16.63±0.91^b	18.24±0.05^a
P4	316.00±7.16^a	284.00±5.29^{ab}	415.63±4.88^a	417.87±29.03^a	14.64±0.21^c	14.87±0.10^c

注：不同小写字母表示各处理间差异显著（$P<0.05$）。
Note: Different lowercase letters indicate significant differences among the different treatments ($P<0.05$).

镉胁迫对甘蔗抗氧化酶系统及非蛋白巯基物质的影响

为了解镉胁迫下甘蔗的生理生化响应，通过桶栽试验模拟甘蔗在镉胁迫下生长，探讨镉胁迫的不同浓度和不同时期对甘蔗生理生化的影响（图10）。结果表明：镉处理对甘蔗叶绿素均有不同程度的影响，随着时间的推移叶绿素变化得更为明显；对于清除活性氧系统酶类，包括超氧化物歧化酶（SOD）和过氧化物酶（POD），相比于对照则出现了较大的波动；脯氨酸（PRO）和丙二醛（MDA）含量受镉浓度影响较小，在镉处理前期并未有明显的变化，但在处理中后期出现了明显的上升趋势；谷胱甘肽（GSH）和非蛋白巯基（NPT）含量在不同的镉浓度和处理时期都有增加的趋势，而植物络合素（PCs）则随着镉浓度的增加会出现先升高后趋于平稳的情况。甘蔗生理不仅受镉浓度处理的影响，随着时间的推移响应的镉积累也是左右甘蔗生理变化的重要因素。

（范业赓，廖洁，王天顺，丘立杭，陈荣发，黄杏，莫磊兴*，吴建明*）

Effects of cadmium stress on antioxidant enzyme system and non-protein thiols substances in sugarcane

In order to understand the physiological and biochemical responses of sugarcane under cadmium stress, barrel experiments were conducted to simulate the growth of sugarcane under cadmium stress, and to explore the effects of different concentrations and different periods of cadmium stress on the physiological and biochemical characteristics of sugarcane (Fig. 10). The results showed that cadmium treatment had different effects on the chlorophyll content of sugarcane, and the change of chlorophyll content was more obvious with the passage of time. For scavenging enzymes of active oxygen system, including superoxide dismutase (SOD) and peroxidase (POD), there was a large fluctuation compared with the control. The contents of proline (PRO) and malondialdehyde (MDA) were less affected by the concentration, and did not change significantly in the early stage of cadmium treatment, but showed an obvious upward trend in the middle and late stage of treatment. The contents of glutathione (GSH) and non-protein thiol (NPT) tended to increase at different cadmium concentrations and treatment periods, while plant complexes (PCs) tended to increase first and then stabilize with the increase of cadmium concentration. Sugarcane physiology is not only affected by cadmium concentration, but also affected by

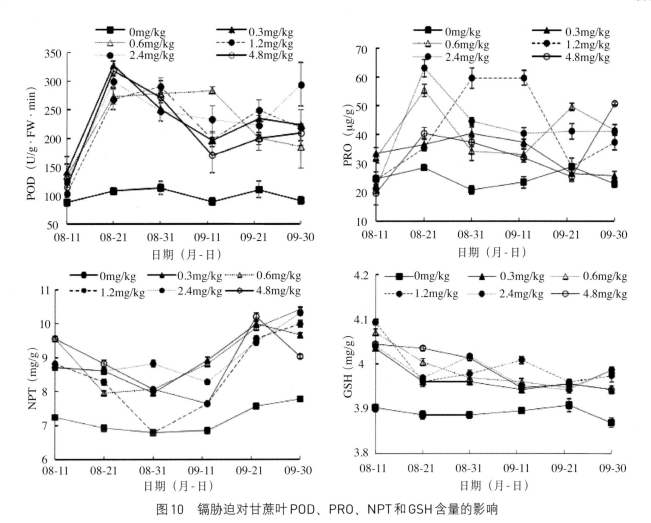

图10 镉胁迫对甘蔗叶POD、PRO、NPT和GSH含量的影响

Fig. 10 Effect of cadmium on POD, PRO, NPT and GSH in leaves of sugarcane leaves cadmium accumulation over time.

(Ye-geng Fan, Jie Liao, Tian-shun Wang, Li-hang Qiu, Rong-fa Chen, Xing Huang, Lei-xing Mo[*], Jian-ming Wu[*])

施氮水平对不同甘蔗品种产量和蔗糖分的影响

试验采用裂区设计，以6个甘蔗栽培品种（桂选B9、桂斐1号、新台糖22号、桂糖42号、桂糖46号和桂糖47号）为主区，以尿素施用量（150 kg/hm^{2}和600 kg/hm^2）为副区，于伸长期施入全部氮肥，测定施氮水平对甘蔗农艺性状及产量和糖分的影响（表18）。研究发现不同氮素水平对甘蔗萌芽出苗影响不大，而分蘖率随施氮量增加而提高；同一品种的株高和茎径在低氮和高氮处理间差异不明显，而除了桂糖42号，其他品种高氮处理有效茎数比低氮处理多；不同品种间的株高、茎径和有效茎数存在明显的差异；试验所有品种甘蔗产量随着尿素施用量在150 kg/hm^{2}到600 kg/hm^{2}升高而增加；桂糖42号的高氮处理蔗糖分高于低氮，其他品种甘蔗蔗糖分均是低氮处理高于高氮处理；氮肥施用增加对增加甘蔗分蘖率、有效茎数和甘蔗产量有明显的效应，但降低甘蔗蔗糖分；不同品种对氮肥的响应不同，与品种的特性有很大关系。因此，生产上应结合甘蔗品种特性，有针对性施用氮肥，促进甘蔗的高效节本栽培。

(范业赓，丘立杭，陈荣发，周慧文，黄杏，卢星高，甘崇琨，吴建明*，李杨瑞*)

Effects of nitrogen application level on yield and sucrose content of different sugarcane cultivars

The experiment was designed with a split zone, and six sugarcane cultivars (GXB9, GF1, ROC22, GT42, GT46 and GT47) were used as the main plot, and the application of urea (150 kg/hm^2 and 600 kg/hm^2) was the secondary plot, and all the nitrogen fertilizers were applied on June 10, 2016. The effects of nitrogen application on the agronomic traits and sugar content were determined (Table 18). The results indicated that nitrogen had little effect on seedling emergence, but it enhanced the tillering; within each cultivar, no significant differences for plant height and stalk diameter between two treatments was observed; Except for GT42, the millable stalks was higher in the high nitrogen treatment than the low nitrogen treatment; cultivar differences were significant for plant height, stalk diameter and millable stalks. The cane yield in all sugarcane cultivars increased with the increased urea rates from 150 kg/hm^2 to 600 kg/hm^2, however, an opposite effect was observed for sucrose content except for the cultivar GT42. In conclusion, increasing nitrogen fertilizer rates increased the tillering rate, millable stalks and cane yield, but decreased the sucrose content in sugarcane. Different response of cultivar to nitrogen suggested take into account the cultivar characteristics in sugarcane production.

(Ye-geng Fan, Li-hang Qiu, Rong-fa Chen, Hui-wen Zhou, Xing Huang, Xing-gao Lu, Chong-kun Gan, Jian-ming Wu*, Yang-rui Li*)

表18 施氮水平对不同甘蔗品种农艺性状的影响

Table 18 Effect of nitrogen treatments on agronomic traits of different sugarcane varieties

处理 Treatments	出苗率（%） Emergence rate	分蘖率（%） Tillering rate	株高（cm） Plant height	茎径（cm） Stalk diameter	有效茎数(条/hm^2) Millable stalk
A1B1	78.60abAB	11.33dB	315.73 bcBCD	2.52 cDE	62 412aA
A1B2	75.79 abcABC	14.68dB	317.08bABC	2.45 cdEF	63 127aA
A2B1	65.09eC	11.51dB	294.39efEF	2.66 bABC	37 082eDE
A2B2	71.93 abcdeABC	14.91dB	300.16 defCDEF	2.63 bBCD	42 879 cdeBCDE
A3B1	70.96 bcdeABC	29.76 abcAB	299.13 defDEF	2.28eG	41 688 cdeCDE
A3B2	64.91eC	29.80 abcAB	289.65fF	2.38 dFG	46 531 bcdBCD
A4B1	74.47 abcdABC	34.49abA	317.71bAB	2.65 bBCD	52 487bB
A4B2	79.65aA	38.04aA	333.50aA	2.53cCDE	51 693bB
A5B1	72.63 abcdeABC	20.39 cdAB	309.63 bcdBCDE	2.78aA	46 452 bcdBCDE
A5B2	71.22 bcdeABC	22.62 bcdAB	310.08 bcdBCDE	2.68 bAB	48 040 bcBC
A6B1	67.63 deBC	11.57dB	304.54 cedBCDEF	2.64 bBCD	36 844eE
A6B2	68.77 cdeABC	15.08dB	315.23 bcBCD	2.70 abAB	40 099 deCDE

注：同列数据后不同大、小写字母分别表示差异达极显著（$P<0.01$）和显著水平（$P<0.05$），下同。

Note: uppercase and lowercase indicate difference is significant at the 0.01 and 0.05 levels, respectively, hereinafter.

不同植物生长调节剂浸种对甘蔗分蘖及产量性状的影响

通过不同植物生长调节剂浸种试验,探讨其对甘蔗萌芽、整齐度、苗期生长、甘蔗分蘖、产量性状及宿根性的影响,为筛选不同化学药剂对延长甘蔗宿根年限的应用提供技术支撑(表19、表20)。以桂选B9为材料,设5个药剂处理,分别为200 mg/L浓度的赤霉素(GA)、细胞分裂素(6-BA)、脱落酸(ABA)、多效唑(PP_{333})、萘乙酸(NAA)溶液和1个对照CK(清水),浸种24 h后种植;调查新植和第一年宿根农艺性状和经济产量等指标。试验结果表明,在5个药剂处理中,新植蔗ABA和NAA处理的甘蔗出苗受到严重抑制,分蘖率则分别为115.6%和98.0%,明显高于CK的97.5%,而有效茎不足(分别为67 260条/hm^2和70 515条/hm^2),低于CK的73 050条/hm^2,因此产量也比对照略低。而GA和6-BA处理发现出苗率分别为41.2%和19.1%,远低于对照82.6%;分蘖率分别为28.9%和43.1%,同样低于对照的97.5%;说明GA和6-BA抑制了甘蔗的出苗和分蘖,导致甘蔗产量显著于对照。PP_{333}处理后甘蔗出苗率为90.8%,高于对照的82.6%,而分蘖率92.3%低于对照97.5%,产量上差异也不明显,在新植蔗中各种性状的表现与对照差异不明显;新植甘蔗品质分析发现PP_{333}处理甘蔗糖分最高,但不同激素对甘蔗品质影响不大。在宿根蔗中发株率、有效茎和茎径均高于对照,产量为111 841.65 kg/hm^2,高于对照的105 435.75 kg/hm^2,说明PP_{333}浸种有利于宿根蔗高产。各处理对甘蔗品质影响不大。PP_{333}浸种处理在新植蔗中能促进甘蔗出苗,在宿根蔗中表现能促进产量增加;PP_{333}浸种处理要优于其他处理,可作为甘蔗浸种剂进一步筛选出最佳使用浓度和使用方法。

(范业赓,陈荣发,周慧文,丘立杭,黄杏,吴建明*,李杨瑞*)

Effects of seed-cane soaking by various plant growth regulators on tillering and yield traits of sugarcane

The effects of soaking cane setts in various plant growth regulators on germination rate, uniformity, seedling growth, tillering rate, yield and ratooning ability of sugarcane were explored, and it is expected to provide technical supports for extending ratoon years by using suitable chemical agent (Table 19、Table 20). The cane setts of sugarcane cultivar Guixuan B9 was used as experimental material, which was soaked in 200 mg/L GA, 6-BA, ABA, paclobutrazol (PP_{333}), NAA and clean water (control) for 24 h, and then planted in the experimental field. The results showed that the germination of cane setts treated by ABA and NAA was severely inhibited among the five treatments, the tillering rates were 115.6% and 98.0%, respectively, which were significantly higher than the control (97.5%), while the millable stalks of ABA and NAA were insufficient (67 260 stalks/hm^2 and 70 515 stalks/hm^2 respectively), which is lower than of the control (73 050 stalks/hm^2), consequently, the yield performed slightly lower than the control. The GA and 6-BA treatments performed much lower emergence rate of 41.2% and 19.1%, respectively, which was much lower the control (82.6%); and performed lower tillering rate of 28.9% and 43.1%, respectively, which was also lower than the control; the inhibition by GA and 6-BA on emergence and tillering were leads to significantly lower cane yield than the control. For the PP_{333} treatment, the emergence rate was 90.8%, performed higher than the control, but the tillering rate was 92.3% and lower than control. As compared with the control, there was no significant difference in cane yield. The sugar content of plant crop was the highest in PP_{333} treatment, however, there was slightly difference among the treatments by various plant growth regulators for juice quality. However, the germination rate, millable stalks and stem diameter were higher than the control in ratoon crop. The cane yield was 111 841.65 kg/hm^2 for PP$_{333}$ treatment, which was significantly higher than the control (105 435.75 kg/hm^2), which indicated the advantage of PP$_{333}$ treatment on ratoon crops. PP$_{333}$ soaking treatment promoted emergence rate in plant crop, and cane yield in ratoon crop; it was the best among all the tested cane setts soaking agencies, further test for screening its suitable application rates is needed.

(Ye-geng Fan, Rong-fa Chen, Hui-wen Zhou, Li-hang Qiu, Xing Huang, Jian-ming Wu[*], Yang-rui Li[*])

表19 不同植物生长调节剂处理对新植甘蔗农艺性状和产量的影响

Table 19 Effects of various plant growth regulators on agronomic traits and yield of sugarcane in plant crop

处理 Treatments	出苗率（%） Emergence rate	分蘖率（%） Tillering rate	株高（cm） Plant height	茎径（mm） Stem diameter	有效茎数（条/hm^2） Millable stalk	产量（kg/hm^2） Yield
GA	41.2bBC	28.9bC	275cB	25.2 cB	60 900bB	63 206.25bB
6-BA	19.1cC	43.1bBC	298bcAB	28.5 aA	33 195cC	54 551.10bB
ABA	52.8bB	115.6aA	328abAB	26.7 bcAB	67 260abAB	88 401.30aA
PP333	90.8aA	92.3aAB	351aA	27.6 abA	70 755aAB	94 555.20aA
NAA	61.5bAB	98.0aAB	335abA	26.6 bcAB	70 515aAB	94 000.20aA
CK	82.6aA	97.5aAB	348aA	27.2 abAB	73 050aA	102 575.85aA

注：同列数据后不同大、小写字母分别表示差异达极显著($P<0.01$)和显著水平($P<0.05$)，下同。

Note: the lowercase and uppercase within a same column represent the difference is significant at 0.05 and 0.01 level. respectively, which means the same hereinafter.

表20 不同植物生长调节剂处理对宿根甘蔗农艺性状和产量的影响

Table 20 Effects of various plant growth regulators on agronomic traits and yield of sugarcane in ratoon crop

处理 Treatments	发株率（%） Sprouting	株高（cm） Plant height	茎径（mm） Stem diameter	有效茎数（条/hm^2） Millable stalk	产量（kg/hm^2） Yield
GA	218.81abA	403aA	25.9aA	67 020aA	96 651.45bA
6-BA	252.12aA	369aA	25.7aA	49 545bB	70 392.30cB
ABA	177.62bB	382aA	25.8 aA	65 190aA	104 838.15abA
PP333	190.42bAB	378aA	26.5 aA	71 385aA	111 841.65aA
NAA	169.14bB	378aA	26.1 aA	65 910aA	105 227.25abA
CK	187.76bAB	381aA	26.4 aA	70 350aA	105 435.75abA

注：数据后不同大、小写字母分别表示差异达极显著（$P<0.01$）和显著水平（$P<0.05$）。

高、低糖甘蔗品种成熟期糖分积累特征及代谢相关酶活性分析

为分析高、低糖甘蔗品种蔗糖代谢相关酶与糖分积累的相关性和差异性，对甘蔗高糖（GT35）和低糖（B8）品种成熟期不同成熟度的叶和茎中的蔗糖合成和分解方向酶活性、蔗糖、葡萄糖和果糖含量进行分析（表21）。结果表明，2个甘蔗品种叶中已糖含量与NI酶活性显著负相关，茎中蔗糖含量与SPS和CIN酶活性显著正相关，而与SS-s、SS-c、SAI和NI酶活性显著负相关。在GT35成熟茎和老茎中蔗糖含量显著高于B8，已糖含量则显著低于B8。GT35成熟叶、

成熟茎和老茎中SS-s酶活性显著高于B8，高SS-s酶活性有利于蔗糖合成。茎中SS-c、SAI和NI酶活性由高到低依次为幼茎、成熟茎和老茎，分解酶活性降低有利于甘蔗茎间蔗糖积累。而随着节间成熟，CIN酶活性提高，有利于茎间蔗糖运输和积累。基于以上结果，认为叶中高SS-s和SPS酶活性，茎中高SPS、SS-s和CIN酶活性，低SAI、SS-c和NI酶活性，可能是高糖甘蔗品种成熟期茎中蔗糖积累的重要调节因素。

（牛俊奇，苗小荣，王道波*，杨丽涛，李杨瑞*）

Relationship between sugar accumulation and sucrose metabolism enzymes in the high and low sugar sugarcane at maturing stage.

To analyze the correlation and difference of sucrose metabolic enzymes and sucrose accumulation in high and low sugar sugarcane, the experiment was conducted with the split plot to analyze the sucrose synthesis and decomposition direction enzyme activity, sucrose, glucose and fructose contents in leaves and stems of different maturity at the maturing stage of high sugar (GT35) and low sugar (B8) sugarcane (Table 21). The results showed that in the GT35 and B8 the content of sucrose in the leaves was positively correlated with the activity of neutral/alkaline invertase (NI), the sucrose content in the stem was positively correlated with the activity of sucrose phosphate synthase (SPS) and cell wall-bound invertase (CIN), and negatively correlated with the activity of the sucrose synthase in the synthesis direction (SS-s), the sucrose synthetase in the cleav-age direction (SS-c), soluble acid invertase (SAI) and NI. The sucrose content in the mature and matured stems of GT35 was significantly higher than that of B8, while the hexose content was significantly lower than that of B8. During the sugarcane maturing stage, the activity of SS-s in the mature leaves, mature stems and matured stems of GT35 was significantly higher than that of B8, and high SS-s activity was beneficial to su-crose synthesis. The activities in SS-c, SAI and NI in sugarcane stems were in the order of immature stem > maturing stem > matured stem, and its enzyme activity reduced to facilitate the sucrose accumulation in the stem. With the increase of CIN enzyme activity during the maturing period of the internode, it is beneficial to the transportation and accumulation of sucrose in sugarcane stems. The analysis demonstrated that high SS-s and SPS activities in leaves, high SPS, SS-s and CIN activities in stems, and low SAI, SS-c and NI activities in stems may be important factors for regulating sucrose accumulation in high sugar varieties of sugarcane at the maturing stage.

(Jun-qi Niu, Xiao-rong Miao, Dao-bo Wang*, Li-tao Yang, Yang-rui Li*)

表21 成熟期高、低糖甘蔗品种叶和茎中蔗糖代谢相关酶活性与糖分含量的相关性分析

Table 21 Relationship between sucrose metabolism and sucrose-metabolizing enzymes activities of high and low sugar sugarcane at maturing stage

相关性 Relationships	品种 Variety	SPS酶活性 SPS activities	SS-s酶活性 SS-s activities	SS-c酶活性 SS-c activities	NI酶活性 NI activities	SAI酶活性 SAI activities	CIN酶活性 CIN activities
叶中蔗糖含量 Sucrose content	B8	0.068	0.876*	0.507	-0.882	0.988**	-0.742
	GT35	0.768	0.714	-0.989**	0.559	0.585	-0.711

(续)

相关性 Relationships	品种 Variety	SPS酶活性 SPS activities	SS-s酶活性 SS-s activities	SS-c酶活性 SS-c activities	NI酶活性 NI activities	SAI酶活性 SAI activities	CIN酶活性 CIN activities
叶中己糖含量 Hexose content	B8	0.146	0.911	0.438	-0.916**	0.973**	0.687
	GT35	0.317	0.392	0.950	-0.976**	-0.968**	-0.397
茎中蔗糖含量 Sucrose content	B8	0.821*	-0.920**	-0.943**	-0.978**	-0.932**	0.998**
	GT35	0.990**	-0.956**	-0.995**	-0.983**	-0.946**	0.862*
茎中己糖含量 Hexose content	B8	-0.705	0.544	0.491	0.372	0.517	-0.188
	GT35	-0.964*	0.911*	0.998*	0.951*	0.897*	-0.791

注：*表示差异达显著水平（$P<0.05$）、**表示差异达极显著水平（$P<0.01$）。
Note: "*" and "**" mean correlation are significant at 0.05 and 0.01 level, respectably.

不同代数健康种苗分蘖差异及浸种处理对一代种茎繁育的影响研究

以桂糖46号为实验材料，研究其不同代数分蘖差异及浸种处理对甘蔗产量及其构成因素的影响，为甘蔗健康种苗大田高效繁育提供参考依据（表22）。以清水处理为对照，通过预备试验观察筛选出的甘蔗浸种剂（GZJ）40 mg/L对甘蔗健康种苗一代种茎进行浸种2 h，分析两者之间的大田繁育差异。试验结果表明，不同代数茎尖脱毒健康种苗分蘖率随着代数增加呈骤减趋势，健康种苗原种显著高于一代种茎和二代种茎，分别提高了317.95%和1 937.64%，一代种茎又显著高于二代种茎，提高了387.53%。浸种处理的单株平均分蘖数比对照提高了25.17%，差异达到极显著水平；甘蔗茎径、有效茎、单茎重分别高于对照1.30%、4.98%、5.15%；增产8.40%；株高比对照低3.33%；差异达到极显著水平。说明健康种苗原种优势强；GZJ药剂主要通过提高分蘖率、促进形成甘蔗有效茎、增大茎粗和提高单茎重达到增产目的。

（周慧文，范业赓，陈荣发，黄杏，杨柳，卢星高，吴建明，丘立杭*，李杨瑞*）

Tillering difference among generations of virus-free sugarcane setts and the impact of soaking treatment on the first cane setts multiplication

Aimed at providing guidance for multiplication of virus-free cane setts, the tillering differences among generations of virus-free cane setts of sugarcane cultivar Guitang46 and the impact of soaking treatment in GZJ, a tiller enhancing solution on the cane yield and yield components were studied (Table 22). The first generation of virus free cane setts from sugarcane cultivar Guitang46 was soaked in a cane setts soaking agent for 2 h, and soaking in clean water was considered as the control. The number of tillers, cane yield and its components were investigated in both treatments. The results showed that the tillering declined with increased vegetative generations, the tillers of original virus free cane setts was much higher by 317.95% and 1 937.64% than the tillers from the first and second generation, respectively, and the first generation increased 387.53% as compared with the second generation. As compared with the control, stalk diameter, millable stalks, single stalk weight were increased by 1.30%, 4.98% and 5.15%, respectively, the cane yield increased by 8.40%, while the plant height was 3.33% shorter. Therefore, the virus-free cane setts was of high superiority. The GZJ seed soaking agent could increase yield by promoting the tillering rate, the number of millable stalks, the stem diameter and single stalk weight.

(Hui-wen Zhou, Ye-geng Fan, Rong-fa Chen, Xing Huang, Liu Yang, Xing-gao Lu, Jian-ming Wu, Li-hang Qiu*, Yang-rui Li*)

表22 不同代数甘蔗健康种苗分蘖率

Table.22 Tillering rale of virus-free sugarcane setts in different generation

处理 Treatments	分蘖率（%）Tillering rater			
	I	II	III	平均 Mean
健康种苗原种 Breeder seed of virus-free sugarcane setts	1 096	1 238	1 206	1 180.00aA
一代种茎 First generation seed setts	287	227	333	282.33bB
二代种茎 Second generation seed setts	56.02	57.22	60.49	57.91cC

5.2.2 甘蔗轻简栽培技术 Simplified Cultivation Techniology for Sugarcane

不同体积切种处理对甘蔗单芽育苗效果的影响

通过切割不同体积单芽茎段和使用ABT溶液浸泡单芽茎段，探讨不同体积单芽茎段处理方式对育苗效果的影响，为推广甘蔗单芽茎段繁殖技术提供技术支撑（表23、表24）。以桂糖42号为供试品种，设3个不同处理，分别为完整单芽茎段，1/2单芽茎段和1/4单芽茎段，其中每个处理又分为清水浸泡茎段2小时与50 mg/L ABT浸泡2小时2个处理。出苗后调查甘蔗苗株高、苗粗、展开叶片数和根鲜重干重等指标。结果表明，完整处理的单芽茎段各项指标均极显著优于1/2单芽茎段和1/4单芽茎段；1/2单芽茎段育苗效果显著优于1/4单芽茎段；而ABT浸泡种茎对育苗效果没有显著影响。

（周慧文，范业赓，陈荣发，丘立杭，黄杏，翁梦苓，周忠凤，吴建明*）

Effects of the stem with single bud treated by different cutting treatments on plantlets growth of sugarcane

By cutting the stem with single bud in different ways and soaking the stem with single bud with ABT solution, the effects of different treatments on plantlets growth were researched and it could provide technical support for popularizing the stem with single bud propagation technology of sugarcane (Table 23 and Table 24). Taken as the experiment materials, three different treatments that were set up, namely, ring-cut the stem with single bud, semi-cut the stem with single bud and sliced-cut the stem with single bud; and each treat-ment was divided into two treatments that soaking stem with water for 2 h and soaking stem with 50 mg/L ABT for 2 h. After planting for 90 days, plantlet height, Stem diameter, leaf number and root fresh weight and dry weight were measure. The results showed that the indices of the stem with single bud treated by ring cutting were prominently higher than that of the stem with single bud treated by semi cutting and the stem with single bud treated by slice cutting. The plantlets growth effect of the stem with single bud treated by semi cutting was significantly better than that of the stem with single bud treated by slice cutting; whether the stem was soaked in ABT or not had no significant effect on plantlets growth.

(Hui-wen Zhou, Ye-geng Fan, Rong-fa Chen, Li-hang Qiu, Xing Huang, Meng-ling Weng, Zhong-feng Zhou, Jian-ming Wu*)

表23　ABT浸泡种茎对育苗的影响

Table 23　Effect of seedcane soaking with ABT on plant growth

处理方法	株高(cm)	苗秆粗(mm)	展开叶片数(片)	根鲜重(g)	根干重(g)
环切(A1)	9.22Aa	4.28Aa	2.5Aa	0.99Aa	0.16Aa
对切(A2)	6.78Bb	3.24Bb	2.0Aab	0.50Bb	0.11Aa
片切(A3)	5.83Bb	3.00Bb	1.67Ab	0.56Bb	0.10Aa

表24　不同芽处理方法和ABT浸泡种茎对育苗的影响差异

Table 24　Growth difference in different treatments and seedcane soaking with ABT

处理方法		株高(cm)	苗秆粗(mm)	展开叶片数(片)	根鲜重(g)	根干重(g)
A1	0	8.95Aa	4.30Aa	3.00 Aa	1.20 Aa	0.22 Aa
	50	9.22Aa	4.28Aa	2.50 Aa	0.99 Aa	0.16 Aa
A2	0	6.70Aa	3.37 Aa	2.00 Aa	0.67 Aa	0.15 Aa
	50	6.78Aa	3.24 Aa	2.00 Aa	0.50 Aa	0.11 Aa
A3	0	6.02Aa	2.85 Aa	1.17 Ab	0.41 Aa	0.07 Aa
	50	5.83Aa	3.00 Aa	1.67 Aa	0.56 Aa	0.10 Aa

甘蔗健康种苗原苗提前移栽对田间繁育的影响

以正常7～10叶龄移栽为对照，以4叶龄提前移栽为处理，比较原种健康种苗生产成本、前期长势、分蘖、有效茎数等性状的差异，为甘蔗健康种苗大田繁育节约生产成本和出苗时间提供新技术（表25）。研究结果表明，甘蔗健康种苗原苗提前移栽法具有较多优点，节约约60%的生产成本，缩短大田移栽提前培育时间30～45 d，群体长势均匀，主苗与分蘖苗差异不明显且数量高于对照，同时，便于运输，但移栽前期对水分要求较高；株高、茎径、有效茎数、蔗茎芽数、单茎重均比对照有所提高，提高幅度在0.60%～6.59%；蔗种产量和扩繁量均显著高于对照，原因是分蘖、株高、蔗茎芽数、有效茎等多方面的提高共同导致这个现象。甘蔗健康种苗原苗提前移栽法综合效果优于常规方法，推广前景巨大，但要保证前期水分供应。

（周慧文，范业赓，黄杏，陈荣发，杨柳，卢星高，吴建明，丘立杭*，李杨瑞*）

Effects of early transplanting on the growth and developments of virus-free sugarcane

Taking normal transplanting (7-10 leaves) as control, the virus-free sugarcane seedlings were transplanted at the age of approximately 4 leaves. The production costs, early growth vigour, tillering, and millable stalks were compared with the control (Table 25). The results indicated a range of advantages of transplanting earlier including saving 60% of production costs, shortening 30-45 d for field transplanting, being convenient for transportation, and no obvious difference among main shoots and tillers, and significantly greater amounts of main shoots and tillers. The only disadvantage was higher requirement for carefully water supply in early growth period. The plant height, stalk diameter, millable stalks, buds per stalk and single stalk weight were greater than the control by a range of 0.60%-6.59%. The final propagated amount and cane yield were significantly higher than the control because of the comprehensively better performance in tillers, plant height, buds per stalk and millable stalks. Transplanting pathogen-free seedlings earlier was superior to normal practices, and should be a promising method given that adequate irrigation is ensured at the crop establishment stage.

(Hui-wen Zhou, Ye-geng Fan, Xing Huang, Rong-fa Chen, Liu Yang, Xing-gao Lu, Jian-ming Wu, Li-hang Qiu[*], Yang-rui Li[*])

表25 不同方法分蘖比较

Table 25 The comparison of tillers between early and conventional transplanting

处理 Treatments	移栽成活率 Transplant survival rate	单株平均分蘖数 Average number of tillers per plant	单株平均有效分蘖 Average effective tiller per plant
提前移栽法 Early transplanting method	98.60% aA	12.95aA	7.44aA
常规移栽法 Conventional transplanting method	99.03% aA	11.80bA	7.07bB
提高/% Increase	-0.43	9.75	5.23

不同种茎处理对甘蔗萌芽和幼苗生长的影响

以不同部位、不同芽数和不同质量的甘蔗种茎为材料，分析不同种茎来源对甘蔗萌芽和幼苗生长的影响（表26、表27）。结果表明：中部种茎萌芽出苗及幼苗生长表现最好，萌芽率和苗高分别为72.81%和46.33 cm；单芽茎的萌芽率较低，苗较矮，而三芽茎、四芽茎和多芽茎的萌芽率和苗高差异不明显。正常种茎的萌芽率与倒伏种茎的差异不显著，与绵蚜危害种茎、螟虫危害种茎和黑穗病危害种茎的差异显著；正常种茎苗高与螟虫危害种茎的差异不显著，与绵蚜危害种茎、倒伏种茎和黑穗病危害种茎的差异显著；而黑穗病危害种茎芽萌发长成的植株只有3.82%为正常植株，其他均为黑穗苗。结果说明生产上选健康的甘蔗中部或梢部茎留种，以双芽段、三芽段、四芽段或多芽段种茎种植有利于甘蔗萌芽和幼苗生长。

（罗亚伟，覃振强，梁阗，王维赞，李德伟）

Effects of different seed cane on germination and emergence of sugarcane

The effects of billet (seed cane) size, bud numbers and quality of seedcane on germination and emergence of sugarcane were observed in the field in this study (Table 26 and Table 27). The results showed that the emergence rate and plant height of crops established with the seed canes (billets) from middle part of sugarcane stalk were 72.81% and 46.33 cm, respectively, better than the other two treatments. Meanwhile, the germination rate of the single bud seedcane setts was lower, and the height of seedling was shorter. There was no significant differences on germination and emergence among the seed cane setts with three, four or more buds. Among the quality of seed cane stems, there was no significant difference on germination between seedcane setts from lodged or erect stalks, while there were significant differences between the normal seedcane and other three treatments. In addition, there was no significant difference on the height of the seedling between normal seed cane and borer-damaged seed cane, while there were significant differences between the normal seed cane and other three treatments. There were only 3.82% plants could grow normally while others were smut-affected shoots when the smut-infected cane was used as seed cane. It is concluded that it would benefit germination and emergence of sugarcane when the middle and top parts of healthy stalk are used. There was no difference in crop establishment and growth between crops established with seedcane with two, three, four or more buds.

(Ya-wei Luo, Zhen-qiang Qin, Tian Liang, Wei-zan Wang, De-wei Li)

表26 不同部位种茎对甘蔗萌芽及幼苗生长的影响

Table 26 Effects of seedcane setts from different parts of cane stalk on the germination and seedling growth

处理 Treatments	2月15日萌芽率（%）Germinate rate on 15 February				3月2日萌芽率（%）Germinate rate on 2 March				5月5日苗高（cm）Height of cane seedings on 5 May			
	I	II	III	平均 Average	I	II	III	平均 Average	I	II	III	平均 Average
基部芽 Basal seed stem	39.50	36.80	38.20	38.16bB	60.50	64.50	63.20	62.72bA	41.90	42.40	44.80	43.03bAB
中部芽 Middle seed stem	47.40	47.40	55.30	50.00aA	75.00	69.70	73.70	72.81aA	46.40	46.00	46.60	46.33aA
梢部芽 Tops seed stem	50.00	47.40	48.70	48.68aA	71.10	64.50	71.10	68.86abA	37.90	40.60	40.20	39.57cB

注：同列数据后不同小写字母表示差异显著（$P<0.05$），不同大写字母表示差异极显著（$P<0.01$）。

Note: Different lowercase letters in the same column indicate significant difference at 0.05 level, different capital letters indicate ex tremely significant difference at 0.01 level.

表27 不同芽数种茎对甘蔗萌芽及幼苗生长的影响

Table 27 Effects of seedcane setts with different bud numbers on the germination and seedling growth

处理 Treatments	2月15日萌芽率（%）Germinate rate on 15 February				3月2日萌芽率（%）Germinate rate on 2 March				5月5日苗高（cm）Height of cane seedings on 5 May			
	I	II	III	平均 Average	I	II	III	平均 Average	I	II	III	平均 Average
单芽茎 Single bud seed stem	13.20	22.40	21.10	18.86bB	51.30	55.30	56.60	54.39bA	29.90	31.40	30.40	30.57cC
双芽茎 Bouble buds seed stem	35.50	39.50	36.80	37.28aA	63.20	65.80	65.80	64.97aA	40.80	38.50	38.50	39.27bB

(续)

处理 Treatments	2月15日萌芽率（%） Germinate rate on 15 February				3月2日萌芽率（%） Germinate rate on 2 March				5月5日苗高（cm） Height of cane seedings on 5 May			
	I	II	III	平均 Average	I	II	III	平均 Average	I	II	III	平均 Average
三芽茎 Three buds seed stem	34.2	31.6	32.9	32.89aAB	64.5	52.6	56.6	57.89abA	42.7	43.0	44.7	43.47aA
四芽茎 Four buds seed stem	47.4	34.2	36.8	39.47aA	63.2	61.8	61.8	62.28aA	43.5	42.3	41.8	42.53aAB
多芽茎 Many buds seed stem	36.8	26.3	31.6	31.58aAB	61.8	64.5	59.2	61.84aA	45.5	43.0	40.7	43.07aAB

注：同列数据后不同小写字母表示差异显著（$P<0.05$），不同大写字母表示差异极显著（$P<0.01$）。

Note: Different lowercase letters in the same column indicate significant difference at 0.05 level, different capital letters indicate ex tremely significant difference at 0.01 level.

一次性施用甘蔗专用缓释肥对甘蔗产量及蔗糖分的影响

研究一次性施用甘蔗缓释肥对甘蔗产量和蔗糖分的影响，旨在解决甘蔗生产中施肥次数多、生产成本高等问题（表28）。以常规施肥（CK）为对照，缓释甘蔗专用肥6个配方肥一次性施用，进行一新一宿2年的栽培试验，调查分析各处理甘蔗的农艺、经济性状及效益。结果表明，与CK相比，缓释肥具有增加甘蔗分蘖的功效，前期甘蔗株高显著增高、茎径增粗；其中，新植蔗施用缓释肥A、B、C、D、E处理产量比CK增产10.27%～15.36%，差异显著；宿根蔗施用缓控释肥A、B、C、D、E、F处理产量比CK增产2.05%～9.28%，但差异不显著。一次性施用甘蔗专用缓释肥，具有提高甘蔗产量的功效；尤其新植蔗一次性施用缓释肥A、B、C、D、E处理，可提高甘蔗分蘖率，增加有效茎数，增产效果显著；宿根蔗一次性施用缓释肥，虽然甘蔗增产效果不显著，但劳力成本显著下降；新植蔗或宿根蔗一次性施用缓释肥中，E处理施肥量2 250 kg/hm^2。（含增效剂8 kg/t）的增产效果和经济效益最佳。

（梁阗，杨尚东，谭宏伟，何为中，卢文，谢金兰，王南通）

Effects of one-time application of special slow-release fertilizer on yield and sucrose content of sugarcane

The effect of one-time application of special slow-release fertilizer (SSRF) on yield and sucrose content of sugarcane was studied in order to solve the problems of high frequency of fertilizing and high production cost in sugarcane production (Table 28). The cultivation experiments (one plant and one ratoon crops) were carried out, in which conventional fertilizer (CK) as control and six SSRF treatments were applied in sugarcane at one time. The agronomic traits, economic characters and benefits of each treatment were investigated and analyzed. The results showed that compared with CK, SSRF had the effect of increasing tiller, and significant increase in plant height and stalk diameter in early growth stage. As compared to the CK, the yields in SSRF treatments A, B, C, D and E significantly increased by 10.27%～15.36% in plant crop and by 2.05%～9.28%. The one-time application of special slow-release fertilizer for sugarcane had positive effect on cane yield. In particular, one-time application of slow-release fertilizer in plant cane increased tillering and the number of millable stalks, and significantly increased cane yield. Even though no significant cane yield increase was observed in ratoon crop, the labor cost saving was significant. The best recommendation for one-time application of slow-release fertilizer was 2 250 kg/hm^2 (Treatment E) with 8 kg/t of synergist.

(Tian Liang, Shang-dong Yang, Hong-wei Tan, Wei-zhong He, Wen Lu, Jin-lan Xie, Nan-tong Wang)

表28　不同施肥处理对甘蔗产量性状及产量的影响

Table 28　Effects of different fertilizer treatments on cane yield and its components

处理 Treatments	有效茎数（条/hm^2）Millable stalk		株高（cm）Plant height		茎径（cm）Stalk diameter		产量（t/hm^2）Cane yield		产量比CK（±%）Yield ratio	
	P	R	P	R	P	R	P	R	P	R
A	68 475abAB	66 000a	327a	350a	2.65a	2.62a	99.87aA	100.34a	14.57	3.43
B	70 425aA	67 080a	336a	349a	2.59a	2.66a	100.38aA	101.00a	15.15	4.11
C	65 070abcAB	64 665a	334a	355a	2.59a	2.65a	96.12aAB	102.00a	10.27	5.14
D	66 870abcAB	67 665a	328a	356a	2.55a	2.64a	97.41aA	100.01a	11.75	3.09
E	67 710abcAB	67 335a	334a	353a	2.62a	2.68a	100.56aA	106.01a	15.36	9.28
F	62 370cB	65 414a	330a	351a	2.61a	2.63a	88.40bB	99.00a	1.41	2.05
CK	63 150bcAB	65 330a	332a	348a	2.54a	2.59a	87.17bB	97.01a	0	0

注：P为新植；R为宿根。

Note: P means Plant crop, R means ratoon crop.

桂辐98-296种茎补种桂糖42号宿根蔗农艺性状与效益分析

研究桂辐98-296（GF98-296）种茎直接补种桂糖42号（GT42）宿根蔗对农艺性状的影响，探讨GF98-296解决甘蔗宿根缺株断垄问题，为推广应用提供科学依据（表29、表30、表31）。试验设计GF98-296种茎直接补种GT42宿根（A）、GT42种茎直接补种于GT42宿根（B）和自然状态GT42宿根（CK）3个处理，在GT42宿根苗期1～4张叶片时期实施补种，并调查各个处理的农艺性状及进行经济效益分析。结果表明：A处理产量达到77.93 t/hm^2，分别比B处理和CK增产14.15、19.79 t/hm^2，增幅分别达22.18%和34.03%，达到极显著差异水平，B处理的产量与CK之间差异不显著；直接补种处理均不影响GT42宿根的发株、成茎、产量和蔗糖分；A处理的GF98-296蔗茎蔗糖分略高于GT42宿根；通过GF98-296种茎直接补种GT42宿根，蔗农增收5 506.6元/hm^2，制糖企业可增加工业产值14 813.2元/hm^2。GF98-296适宜作为GT42宿根的种茎直接补种品种，应对这项新技术进行大面积示范和推广。

(游建华，梁阗，樊保宁，吴凯朝，黄日宏，谭宏伟，廖庆才)

Agronomic and beneficial analyses for supplementary planting with GF98-296 seedcanes in GT42 ratoon crop

We studied the effects of supplementary planting for gap filling with GF98-296 seedcane in GT42 ratoon crop on agronomic traits of sugarcane, aiming to solve the problem of plant missing in sugarcane ratoon crops (Table 29, Table 30 and Table 31). Three treatments were designed, which were supplementary planting with seedcane of GF98-296 in GT42 ratoon (A), supplementary planting with seedcane of GT42 in GT42 ratoon (B) and natural ratoon of GT42 (CK). Seedcane supplementary planting was carried out in 1-4 leaf stage of GT42 ratoon crop, the agronomic traits of all treatments were investigated and the economic benefits were analyzed at harvest stage. The results showed that the yield of treatment A was 77.93 t/hm^2, increased by 14.15 t/hm^2 or 22.18% and 19.79 t/hm^2 or 34.03% respectively compared with of treatment B and CK. The difference between the yield of treatment B and CK was not significant. Supplementary planting treatment did not affect the growth, effective stem, yield and sucrose content of GT42. The sucrose content of GF98-296 was slightly higher than that of GT42 in treatment A. Supplementary planting with GF98-296 seedcane in GT42 ratoon crop could increase the income by 5 506.6 yuan/hm^2 for sugarcane farmers, increase industrial output by 14 813.2 yuan/hm^2 for sugar enterprise. GF98-296 is suitable for supplementary planting in GT42 ratoon crop, and this new technology should be demonstrated and extended.

(Jian-hua You, Tian Liang, Bao-ning Fan, Kai-chao Wu, Ri-hong Huang, Hong-wei Tan, Qing-cai Liao)

表29 不同处理的产量构成因素表现

Table 29 Performances of yield components in different treatments

处理	品种及植期	有效茎数 (条/hm^2)	茎径 (cm)	株高 (cm)	产量 (t/hm^2)	单茎重 (kg/条)	合计产量 (t/hm^2)	比CK (t/hm^2)	比CK (%)
A	GF98-296补种	18 720	2.41	281	21.34	1.14	77.93Aa	19.79	34.03
	GT42宿根蔗	43 770	2.56	272	56.59	1.29			
B	GT42补种	6 160	1.92	171	5.85	0.95	63.78Bb	5.64	9.70
	GT42宿根蔗	44 515	2.53	271	57.93	1.30			
CK	GT42宿根蔗	45 765	2.58	265	58.14	1.27	58.14Bb	0	—

表30 不同处理的农业经济效益

Table 30 The agricultural economic benefits in different treatments

处理	增产蔗量 (t/hm^2)	农业产值 (元/hm^2)	补种量 (t/hm^2)	补种系列成本(元/hm^2)			成本小计 (元/hm^2)	纯增收 (元/hm^2)
				蔗种成本	补种人工	砍收人工		
A	19.79	9 895.0	2.34	1 113.6	900.0	2 374.8	4 388.4	5 506.6
B	5.64	2 820.0	2.55	1 224.0	900.0	676.8	2 800.8	19.2
CK	0	0	0	0	0	0	0	0

表31　2017年试验结果工业经济效益分析

Table 31　Analysis of industrial economic benefits in 2017

处理	品种与植期	单产 (t/hm²)	蔗糖分 (%)	产糖量 (t/hm²)	合计 (t/hm²)	产值 (元/hm²)	比CK增 (元/hm²)	A比B增 (元/hm²)
A	GF98-296 补种	21.34	15.3	2.775	9.94	57 663.6	14 813.2	10 979.4
	GT42 宿根蔗	56.59	14.9	7.167				
B	GT42 补种	5.85	14.63	0.727	8.049	46 684.2	3 833.8	—
	GT42 宿根蔗	57.93	14.87	7.322				
CK	GT42 宿根蔗	58.14	14.95	7.388	7.388	42 850.4	—	—

钾肥施用量对甘蔗产量、糖分积累及其抗逆性的效应研究

通过田间试验研究不同钾肥施用量对甘蔗产量、糖分积累及其抗逆性的影响（表32）。结果表明，甘蔗产量随着钾肥施用量增加而增加，4个钾用量处理的新植、宿根两年甘蔗平均产量增产率分别为16.9%、21.8%、25.0%、25.6%。扣除化肥投资K1、K2、K3、K4、CK的收益分别为30 583元/hm²、31 140元/hm²、31 190元/hm²、30 463元/hm²、26 485元/hm²。钾肥施用对甘蔗糖分积累的促进作用不明显，而对甘蔗抗黑穗病有一定作用，宿根甘蔗生长后期黑穗病发病率随着钾肥施用量增加而下降。钾肥施用对提高甘蔗螟虫危害抗性作用不明显。钾肥施用有利于提高甘蔗田间抗旱性，在天气持续干旱条件下，甘蔗保持的青叶数随着钾肥施用量增加而增加。综合各项调查结果，本试验钾肥施用量以 K_2O 300 kg/hm² 表现最优。

（谢金兰，李长宁，李毅杰，梁强，刘晓燕，罗霆，林丽，梁阗，何为中，谭宏伟）

Effects of potassium fertilizer application amount on sugarcane yield, sugar accumulation and stress resistance.

A field experiment was conducted to research effects of potassium fertilizer application amount on sugarcane yield, sugar accumulation and stress resistance (Table 32). The results showed that the yield of sugarcane increased with the increase of potassium fertilizer application amount. The average product of new-planting sugarcane and ratooning sugarcane in two years were increased by 16.9%, 21.8%, 25.0% and 25.6%, respectively. Earnings excluding fertilizer investment were 30 583 yuan/hm², 31 140 yuan/hm², 31 190 yuan/hm², 30 463 yuan/hm² and 26 485 yuan/hm² (control). The effect of potassium fertilizer on sugar accumulation was not significant but had a certain effect on sugarcane smut resistance. The incidence of smut at later stages of ratooning sugarcane decreased with the increase of potassium application rate. The application of potassium fertilizer had no significant effect on improving the sugarcane resistance of borers. Application of potassium fertilizer was beneficial to improve the drought resistance of sugarcane in field. The number of green leaves of sugarcane increased with the increase of potassium application under the condition of continuous drought. K_2O 300 kg/hm² was the best application rate in the experiment.

(Jin-lan Xie, Chang-ning Li, Yi-jie Li, Qiang Liang, Xiao-yan Liu, Ting Luo, Li Lin, Tian Liang, Wei-zhong He, Hong-wei Tan)

表32 新植蔗农艺性状及产量

Table 32 Analysis of agronomic characters and yield of newly planted sugarcane

处理	萌芽率（%）	分蘖率（%）	株高（cm）	茎径（cm）	有效茎（株/hm²）	产量（kg/hm²）
K1	52.4±5.2a	91.7±8.2b	203±9.1ab	2.70±0.04b	63 015±1 175a	67 910±2 102b
K2	51.7±6.9a	102.6±1.3ab	211±3.5a	2.73±0.04ab	64 050±910a	71 350±1 602ab
K3	50.0±4.5a	115.4±10.9a	208±8.1a	2.77±0.04ab	63 335±437a	71 495±1 289ab
K4	48.6±6.2a	115.3±12.4a	203±8.1ab	2.80±0.02a	64 845±806a	72 665±1 725a
CK	49.2±2.7a	98.9±9.9b	198±10.0b	2.70±0.02b	60 080±1 036b	59 210±2 126c

甘蔗-绿豆间作压青还田和施氮水平对甘蔗性状的影响

探讨甘蔗 Saccharum officinarum-绿豆 Vigna radiata 间作和不同施氮水平对甘蔗生长、产量及氮素营养的影响，为甘蔗合理间作提供参考依据（表33）。试验设计3种种植方式（绿豆单作、甘蔗单作、甘蔗-绿豆间作压青还田）和3个施氮水平（不施氮、减量施氮、常规施氮），测定甘蔗不同时期的生长性状。种植方式和施氮水平都显著影响甘蔗的分蘖数、干物质量、氮素吸收量、有效茎数和蔗茎产量；种植方式显著影响甘蔗的出苗数；施氮水平×种植方式显著影响甘蔗的有效茎数、成茎率、收获期干物质量和氮素吸收量。与甘蔗单作处理相比，间作处理使甘蔗出苗数和分蘖数分别降低了9.61%～10.52%和10.30%～11.05%，使有效茎数、干物质量、氮素吸收量和蔗茎产量分别提高了0.15%～14.28%、14.28%～34.76%、24.00%～29.58%和15.88%～20.16%。对于间作处理，甘蔗生长80 d的土地当量比为1.47～1.53，甘蔗收获期的土地当量比为1.76～1.94，甘蔗的竞争能力大于绿豆。与常规施氮的单作甘蔗相比，减量施氮的间作处理不会降低甘蔗的蔗茎产量和土壤氮素营养。甘蔗-绿豆间作处理能提高土地当量比和土壤氮含量，促进甘蔗生长，提高甘蔗产量和氮素吸收。

（苏利荣，何铁光，苏天明，李琴，秦芳，李杨瑞*）

Effects of sugarcane-mungbean intercropping, bean straw returning and nitrogen application level on sugarcane traits

To explore the effect of sugarcane (*Saccharum officinarum* inter-specific hybrids) - mungbean (*Vigna radiata*) intercropping and different nitrogen application levels on sugarcane growth, yield and nitrogen nutrition, and provide a reference for rational sugarcane intercropping (Table 33). Three cropping patterns (monocropping of mungbean, monocropping of sugarcane, intercropping of sugarcane and mungbean with mungbean straw returning), and three nitrogen treatments (no N application, reduced N application, conventional N application) were used in the experiments. Sugarcane traits during different growth period were measured. Tiller number, dry biomass, nitrogen uptake, number of millable stalks and cane yield of sugarcane were significantly affected by nitrogen level and cropping pattern. Sugarcane germination and crop establishment was also significantly affected by cropping system. Number and percentage of millable stalks, dry biomass and nitrogen uptake of sugarcane were significantly affected by nitrogen level×cropping system. Compared with monocropping of sugarcane, intercropping treatment reduced emergence and tiller nubmer by 9.61% ~10.52% and 10.30% ~11.05% respectively, while increased number of millable stalks, dry biomass, nitrogen uptake and cane yield by 0.15%~14.28%, 14.28%~34.76%, 24.00%~29.58% and 15.88%~20.16%, respectively. For the intercropping treatment, the land equivalent ratio was 1.47~1.53 after sugarcane grew for 80 days, and the land equivalent ratio at sugarcane harvest was 1.76~1.94. The competition ability of sugarcane was greater than that of mungbean. Compared with monocropping of sugarcane with conventional N application, intercropping treatment with reduced N application did not decrease cane yield and soil nitrogen level. Intercropping of sugarcane and mungbean can increase the land equivalent ratio and soil nitrogen level, promote sugarcane growth and increase cane yield and nitrogen uptake.

(Li-rong Su, Tie-guang He, Tian-ming Su, Qin Li, Fang Qin, Yang-rui Li[*])

表33 不同种植模式及施氮水平下的甘蔗农艺性状

Table 33 Agronomic traits of sugarcane under different cropping patterns and nitrogen application levels

种植模式 Cropping pattern(C)	施氮水平 (kg/hm^2) Nitrogen level(N)	出苗数 (×10^4/hm^2) Emergency number	分蘖数 (×10^4/hm^2) Tiller number	有效茎数 (×10^4/hm^2) Number of Millable stalks	成茎率（%） Percentage of millable stalks	氮素吸收量 (kg/hm^2) Nitrogen uptake	蔗茎产量 (t/hm^2) Cane yield
甘蔗 Sugarcane	0	9.70±0.12a	8.87±0.13c	4.87±0.05c	26.21±0.37c	64.06±1.61e	55.98±2.18e
	231	9.63±0.17a	12.11±0.13a	5.76±0.08b	26.51±0.58c	100.83±3.44c	83.28±2.34c
	330	9.68±0.10a	12.13±0.12a	6.83±0.09a	31.32±0.44b	143.03±3.76b	101.65±3.32b
甘蔗-绿豆 Sugarcane-mungbean	0	8.68±0.11b	7.89±0.15d	5.43±0.04b	32.80±0.67ab	84.29±5.74d	70.12±1.84d
	231	8.70±0.18b	10.79±0.10b	6.72±0.12a	34.49±0.44a	143.18±2.36b	99.00±1.94b
	330	8.75±0.22b	10.88±0.09b	6.84±0.26a	34.81±1.16a	188.61±5.07a	122.44±3.64a
F	C	55.51**	139.31**	23.41*	123.65**	125.89**	61.62**
	N	0.04	425.98**	86.50**	15.30**	271.26**	174.21**
	C×N	0.06	1.12	6.85*	5.98*	6.14*	0.87

间种大豆对宿根甘蔗的影响

文章研究以B8、新台糖22号（ROC22）和桂糖21号（GT21）三个甘蔗品种的宿根蔗与大豆间作，研究间作大豆对宿根甘蔗生长及性状的影响（表34）。结果表明，宿根蔗与大豆间作对甘蔗的株高、茎径、单茎重没有显著影响，但显著增加甘蔗有效茎数，提高甘蔗产量，使单位面积蔗田的蔗-豆总经济收益比甘蔗单作大幅度增加。因此，甘蔗与大豆间种充分利用了多种农业资源，可以增加单位面积蔗田的经济产出，具有较好的经济效益。

(覃刘东，杨建波，彭东海，杨丽涛[*]，李杨瑞[*])

Effects of intercropping soybean on growth and yield in ratoon sugarcane

In the present study, the ratoon crops of three sugarcane varieties B8, ROC22 and GT21 were cultivated with soybean intercrop (Table 34). The results showed that sugarcane intercropping with soybean had no significant effects on plant height, stalk diameter and single stalk weight, but significantly increased the millable stalks and cane yield, and the total economic benefits compared with the monoculture of sugarcane. Therefore, intercropping sugarcane with soybean is good for full utilization of land, space and energy, and increasing economic benefit per unit area of land.

(Liu-dong Qin, Jian-bo Yang, Dong-hai Peng, Li-tao Yang[*], Yang-rui Li[*])

表34 宿根蔗不同处理的甘蔗产量性状和锤度表现

Table 34 Performances of cane yield components and Brix in ratoon crop with different treatments

处理	株高（cm）	茎径（cm）	单茎重（kg）	有效茎数（条/hm²）	锤度（°Brix）
ROC22间作	347ab	2.75b	1.44bc	65 505b	16.58c
ROC22对照	335ab	2.65bc	1.33c	57 176d	17.24b
B8间作	357a	2.90a	1.65a	66 234b	17.52a
B8对照	342ab	2.87a	1.63a	61 384c	17.35ab
GT21间作	321bc	2.85ab	1.43bc	74 956a	17.76a
GT21对照	307c	2.67bc	1.34c	57 224d	17.22b

5.2.3 甘蔗机械化研究 Sugarcane Mechanization Research

机械化生产对桂糖47号宿根能力的影响与分析

探讨机械收获对桂糖47号宿根能力的影响，为桂糖47号的机械化生产提供依据（表35）。在对桂糖47号新植蔗进行机械化种植和管理的基础上，分别进行机械收获和人工砍收对比试验，连续2年调查宿根蔗的发株率、产蔗量、蔗糖分及相关农艺性状并进行分析。桂糖47号在机械收获后的第1和2年宿根的发株率、株高、茎径、有效茎、产蔗量、蔗糖分与人工收获的差异均不显著。其中机械收获后的第2年宿根产蔗量达到101.70 t/hm²，含糖量达到15.32 t/hm²，与人工砍收的相当。桂糖47号宿根能力强，抗倒能力强，耐机收碾压能力强，适合全程机械化生产，在劳动力缺乏的蔗区应加速推广应用，以降低成本，提高效益。

（王伦旺，邓宇驰，谭芳，唐仕云，黄海荣，经艳，杨荣仲）

Effect of mechanized production on ratooning ability of Guitang 47

To provide a basis for the mechanized production of Guitang 47, the effect of mechanical harvesting on the ratooning ability of Guitang 47 was investigated (Table 35). On the basis of mechanized planting and management of new planting of Guitang 47, mechanical harvesting and manual harvesting were compared from the shooting rate of ratoon sugarcane, cane yield, sucrose content and related agronomic traits continuously for two years. There were no significant differences in plant generation rate of ratoon sugarcane Guitang 47, plant height, stalk diameter, effective stalks, cane yield and sucrose between machine harvesting and manual harvesting, for the 1-year and 2-year ratoon plants. Especially, the 2-year ratoon plants harvested mechanically showed the cane yield of 101.70 t/hm² and the sugar content up to 15.32 t/hm², which were equivalent to those of manual harvesting. Guitang 47 has strong ratooning ablity and strong resistance to lodging, and is adaptive to full mechanized production. It should be popularied and applied in the machine harvesting sugarcane areas where labor in lacking, so as to reduce costs and improve efficiency.

(Lun-wang Wang, Yu-chi Deng, Fang Tan, Shi-yun Tang, Hai-rong Huang, Yan Jing, Rong-zhong Yang)

表35　机收获对桂糖47号宿根蔗发株和分蘖的影响
Table 35　Effects of machine harvesting on ratooning and tillering of Guitang 47

宿根年份 Ratoon age	收获方式 Harvesting method	上茬蔗有效茎数 （千条/hm²） Number of millable canes of preceding crop	发株数 （千条/hm²） Plant number	发株率 （%） Ratooning rate	最高苗数 （千条/hm²） Maximum seeding number	分蘖率 （%） Tillering rate
第1年（2017年） First year (2017)	机收	79.74±3.15aA	106.13±4.56aA	133.1	165.80±5.19aA	56.2
	人工	78.02±2.13aA	106.56±3.16aA	136.6	166.55±6.58aA	56.3
第2年（2018年） Second year (2018)	机收	72.89±2.63aA	102.23±2.91aA	140.3	161.62±3.11aA	58.1
	人工	72.02±1.65aA	103.64±4.45aA	143.9	164.27±5.77aA	58.5

注：表中数据为平均值±标准误。同一指标不同小写字母和大写字母分别表示差异显著（$P<0.05$）和差异极显著（$P<0.01$），下同。

Note: Values are mean ± SE. The same indicalor with different lowercase letters and capital letters indicate significan difference ($P<0.05$) and extremely signifcant difference ($P<0.01$), respectively. The same as below.

植保无人机在蔗田化学除草上的应用效果

蔗田杂草对甘蔗产量有重要影响，为了探讨植保无人机在蔗田化学除草上的应用效果，用植保无人机（3WWDZ-10A）喷施900 g/L 乙草胺乳油进行苗前封闭除草，喷施15% 硝磺草酮悬浮剂与38% 莠去津悬浮剂混配液进行苗后除草，调查杂草防除效果（表36）。结果表明，应用植保无人机飞喷乙草胺对单子叶杂草株防效与鲜重防效分别为88.46%、96.68%，双子叶杂草株防效与鲜重防效分别为90.16%、97.99%，总杂草的株防效和鲜重防效分别为89.66%、97.85%；飞喷硝磺草酮与莠去津混配液对单子叶杂草的株防效与鲜重防效分别为89.29%、86.31%，对双子叶杂草的株防效与鲜重防效分别为99.23%、97.03%，对总杂草的株防效与鲜重防效分别为96.84%、96.25%，防除效率49.95 min/hm²。说明植保无人机进行蔗田杂草具有防除效果好、对甘蔗安全和作业效率高的优势。

（张小秋，宋修鹏，梁永检，宋奇琦，覃振强，李杨瑞[*]，吴建明[*]）

Application of unmanned aerial vehicle on chemical weed control in sugarcane field

Weeds in sugarcane field have great effects on sugarcane yield. In order to explore the control effects of unmanned aerial vehicle (UAV) spaying chemical herbicide in sugarcane field, the 3WWDZ-10A UAV was used to spray 900 g/L acetochlor (emulsifiable concentrate) before germination. The mixture of 15 % mesotrione (suspension concentrate) and 38 % atrazine (suspension concentrate) was sprayed by UAV after sugarcane germination (Table 36). After that, the control effects were investigated. The results showed that the control effects of acetochlor on plant control effect and fresh weight of monocotyledonous weed were 88.46% and 96.68%, that of dicotyledonous weeds were 90.16% and 97.99%, and that of total weed were 89.66% and 97.85%, respectively. The control effects of the mixture of mesotrione and atrazine on plant control effect and fresh weight of monocotyledonous weeds were 89.29% and 86.31%, that of dicotyledonous weed were 99.23% and 97.03%, and that of total weeds were 96.84% and 96.25%, respectively. The control efficiency of UAV was 49.95 min/hm^2. The applications of UAV to conduct chemical weed control had the superiorities of good control effects, high working efficiency and no damages to sugarcane.

(Xiao-qiu Zhang, Xiu-peng Song, Yong-jian Liang, Qi-qi Song, Zhen-qiang Qin, Yang-rui Li[*], Jian-ming Wu[*])

表36 植保无人机蔗田苗前封闭除草效果

Table 36 Control effects of herbicide sprayed by UAV before germination

处理 Treatments	杂草类别 Weed categories	株数（株） Weed number	株防效（%） Control effects of weed number	鲜重（g） Fresh weight	鲜重防效（%） Control effects of fresh weight
施药	单子叶杂草	0.60±0.55*	88.46	0.76±0.70*	96.68
	双子叶杂草	1.20±0.89*	90.16	3.67±2.73*	97.99
	总草	1.80±0.45*	89.66	4.43±2.31*	97.85
对照	单子叶杂草	5.20±1.48	—	22.86±6.60	—
	双子叶杂草	12.20±1.48	—	182.96±29.17	—
	总草	17.40±2.88	—	205.81±28.84	—

5.3 甘蔗功能基因组学研究
Functional Genomics of Sugarcane

5.3.1 组学研究 Genomic

低温胁迫处理对甘蔗转录因子表达的影响

为探明低温胁迫处理对不同甘蔗转录因子表达的影响，以桂糖08-1180和ROC22为试验材料，进行低温和常温处理，经转录组测序后，对转录因子基因的差异表达进行了分析（图11）。结果表明：测序共获得360 404条Transcripts，183 515个Unigenes。低温胁迫下桂糖08-1180获得了566个差异表达转录因子基因，隶属于61类转录因子家族。ROC22获得了769个差异表达转录因子基因，隶属于71类转录因子家族。2个品种的差异表达转录因子基因表现为上调表达基因数多于下调表达基因数。ROC22在MYB、AP2-EREBP、GRAS、WRKY、NAC、Orphans、bZIP、FAR1、C3H、C2H2等家族中具有较多的转录因子基因，而桂糖08-1180在AP2-EREBP、WRKY、MYB、Orphans、bZIP、NAC、GRAS、C2H2、FAR1、bHLH等家族中具有较多转录因子基因。有60个转录因子家族包含了2个品种的差异表达转录因子基因，而BES1、CAMTA、CPP等仅含有ROC22的差异表达转录因子基因，CSD仅含有桂糖08-1180的差异表达转录因子基因。从差异表达倍数来看，2个品种中均以表达倍数在10倍以下的转录因子基因最多。在ROC22中，差异倍数较大的转录因子主要分布在AP2-EREBP、Tify和WRKY家族，在桂糖08-1180中，差异倍数较大的转录因子主要分布在MYB、WRKY和AP2-EREBP家族。本研究为甘蔗抗寒相关转录因子的挖掘及甘蔗抗寒性机理研究提供帮助。

（唐仕云，杨丽涛[*]，李杨瑞，杨荣仲，王伦旺，吴建明）

Effect of low temperature stress treatment on expression of transcription factors in sugarcane

In order to find out the effect of cold stress treatment on expression of transcription factor in different sugarcane, both GT08-1180 and ROC22 under cold stress treatment and normal temperature treatment were sequenced with transcriptome sequencing technologies, and different expression of transcription factor genes were analyzed (Fig. 11). The results showed that 360 404 transcripts and 183 515 unigenes were obtained, and 566 differentially expressed transcription factors (TFs) belonging to 61 families were found in GT08-1180, 769 differentially expressed TFs belonging to 71 families were found in ROC22. In addition, there were more up-regulated TFs than down-regulated TFs in both GT08-1180 and ROC22. MYB, AP2-EREBP, GRAS, WRKY, NAC, Orphans, bZIP, FAR1, C3H and C2H2 were the largest number families of differentially expressed trancription factors in ROC22, then AP2-EREBP, WRKY, MYB, Orphans, bZIP, NAC, GRAS, C2H2, FAR1 and bHLH were the largest number families of differentially expressed trancription factors in GT08-1180. Total 60 trancription factor families had the same differentially expressed trancription factors genes in both GT08-1180 and ROC22, while BES1, CAMTA, CPP, DBP, HRT, LIM, PLATZ, Rcd1-like, S1Fa-like, TIG and VOZ had the special differentially expressed transcription factors genes only in ROC22, CSD had the special differentially expressed trancription factors genes only in GT08-1180. Fold change (FC) values of the most TFs were less than 10 in 2 cultivars. It was found that AP2-EREBP, Tify and WRKY had higher FC values in ROC22, but MYB, WRKY and AP2-EREBP had higher FC values in GT08-1180. In this paper, it is provided the foundation in transcription factors of cold resistance and the mechanism of cold resistance in sugarcane.

(Shi-yun Tang, Li-tao Yang*, Yang-rui Li, Rong-zhong Yang, Lun-wang Wang, Jian-ming Wu)

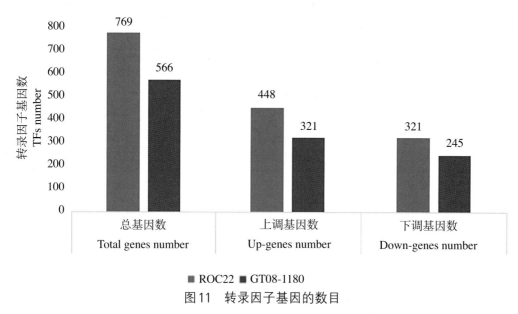

图11 转录因子基因的数目

Fig. 11 The total number of transcription factor genes

Comparative transcriptome profiling of resistant and susceptible sugarcane genotypes in response to the airborne pathogen *Fusarium verticillioides*

Fusarium verticillioides is the pathogen associated with pokkah boeng disease (PBD), the most significant airborne disease of sugarcane (Fig. 12). The molecular mechanisms that regulate the defense responses of sugarcane towards this fungus are not yet fully known. Samples of 'YT 94/128' (resistant, R) and 'GT 37' (susceptible, S) inoculated with *F. verticillioides* on the 14 days post-inoculation were used to analyze the transcriptome to screen R genes. In total, 80.93 Gb of data and 76 175 Unigenes were obtained after assembling the sequencing data, and comparisons of Unigenes with NR, Swiss-prot, KOG, and KEGG databases confirmed 42 451 Unigenes. The analysis of differentially expression genes (DEGs) in each sample revealed 9 092 DEGs in 'YT 94/128,' including 8 131 up-regulated DEGs and 961 down-regulated DEGs; there were 9 829 DEGs in 'GT 37,' including 7 552 up-regulated DEGs and 2 277 down-regulated DEGs. The identified DEGs were mainly involved in catalytic enzyme activity, cell protease, hydrolytic enzymes, peptide enzyme, protein metabolism process of negative regulation, phenylpropanoid metabolism, extracellular region, aldehyde dehydrogenase,

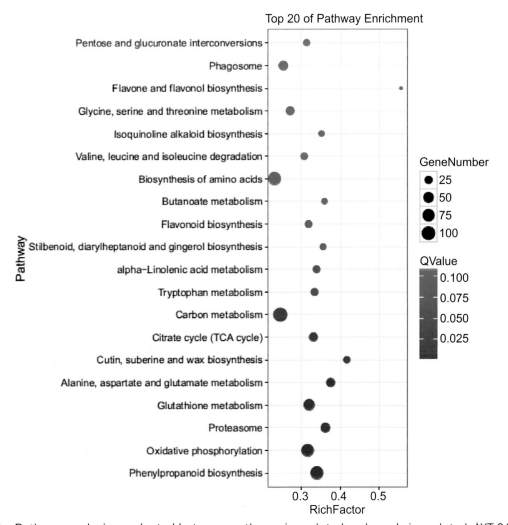

Fig. 12 Pathway analysis conducted between pathogeninoculated and mock-inoculated 'YT 94/128'

endopeptidase, REDOX enzyme, protein kinases, and phosphoric acid transferase categories. KEGG pathway clustering analysis showed that the DEGs involved in resistance were significantly related to metabolic pathways of phenylpropanoid biosynthesis, cutin, suberine and wax biosynthesis, nitrogenous metabolism, biosynthesis of secondary metabolites, and plant-pathogen interactions. This application of transcriptomic data clarifies the mechanism of interactions between sugarcane and *F. verticillioides*, which can help to reveal disease-related metabolic pathways, molecular regulatory networks, and key genes involved in sugarcane responses to *F. verticillioides*.

(Ze-ping Wang, Yi-jie Li, Chang-ning Li, Xiu-peng Song, Jing-chao Lei, Yi-jing Gao, Qiang Liang[*])

甘蔗响应梢腐病菌侵染的蛋白质组学分析

本文旨在解析甘蔗响应梢腐病菌侵染过程中的生理生化机制，为甘蔗抗梢腐病育种及病害防治提供理论指导和科学参考（图13、图14）。以我国甘蔗梢腐病主要致病菌 *Fusarium verticillioides* 孢子悬浮液为病原，以高抗梢腐病品种YT94/128和高感梢腐病品种GT37为宿主材料，在温室条件下进行针刺法接种，提取病情指数最严重时期，即接种后第14天的甘蔗叶片蛋白质进行iTRAQ定量表达分析。结果显示，从GT37中成功鉴定3 707个蛋白，获得542个差异表达蛋白，其中上调表达187个，下调表达355个；从YT94/128中成功鉴定到3 068个蛋白，获得差异蛋白449个，其中上调表达191个，下调表达258个。这说明遗传背景不同的甘蔗品种在蛋白质组成上有很大差异，推测这是不同甘蔗品种间抗梢腐病性差异的重要分子基础。

（王泽平，林善海，梁强，李长宁，宋修鹏，刘璐，李毅杰[*]）

Proteomic analysis of different sugarcane genotypes in response to pokkah boeng disease

This study was to provide a theoretical guidance and scientific reference for sugarcane resistance breeding and disease prevention by analyzing the physiological and biochemical mechanisms in the infection process of sugarcane response to pokkah boeng disease (Fig. 13 and Fig. 14). Using spore suspension of *Fusarium verticillioides* (the main pathogenic fungus of pokkah boeng disease in China) and the reaction of YT94/128 (HR) and GT37 (HS) to *F. verticillioides* wastested by syringe inoculation in greenhouse, and the samples were collected at the 14th day post-inoculation with the highest disease index to carry out quantitative expression analysis by isobaric tags for relative and absolute quantitation (iTRAQ). The results showed a total of 3 707 proteins were successfully identified from GT37, and 542 differentially expressed proteins (DEPs) were obtained, of which 187 were up-regulated and 355 were down-regulated. Meanwhile, a total of 3 068 proteins were successfully identified from YT94/128, and 449 DEPs were obtained, of which 191 were up-regulated and 258 were down-regulated. In brief, there were significant differences in the protein composition of sugarcane varieties with different genetic backgrounds, and it is speculated that this important molecular differences observed in this study explain the resistance variation for pokkah boeng disease between different sugarcane varieties.

(Ze-ping Wang, Shan-hai Lin, Qiang Liang, Chang-ning Li, Xiu-peng Song, Lu Liu, Yi-jie Li[*])

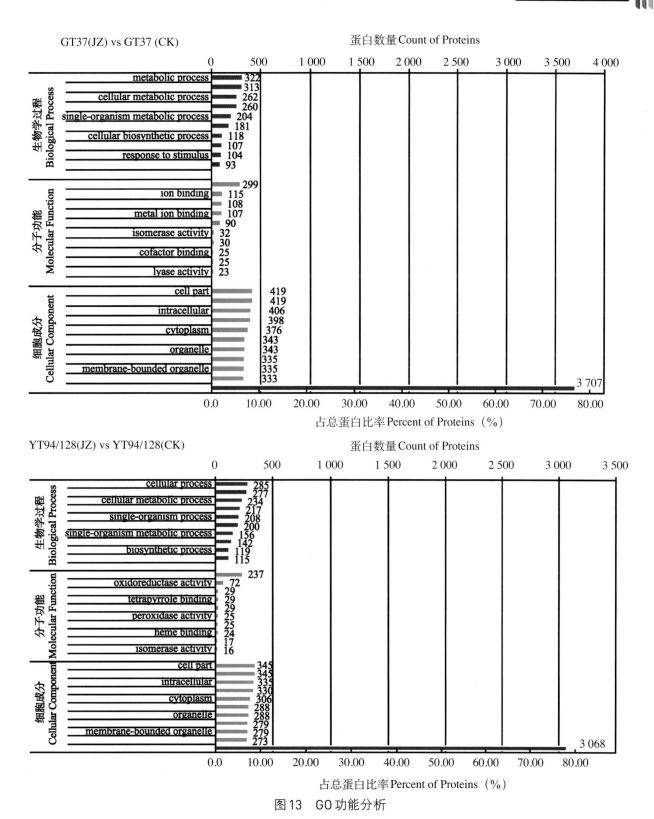

图 13 GO 功能分析

Fig. 13 GO function analysis

注：CK、JZ 分别表示采用清水和孢子悬浮液接种第 14 天的样品。

Note: CK and JZ represent samples at 14th day with syringe inoculation by water and spore suspension, respectively.

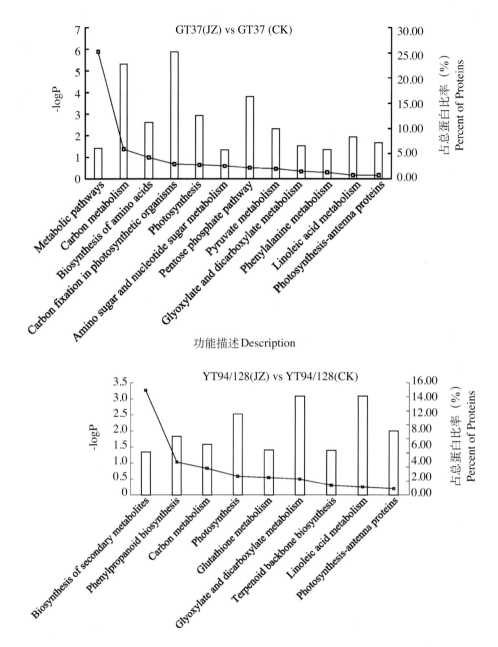

图14 差异表达蛋白 Pathway 显著性富集分析

Fig. 14 Pathway analysis of differential expressed proteins

CK、JZ 分别表示采用清水和孢子悬浮液接种第14天的样品。

CK and JZ represent samples at 14th day with syringe noculation by water and spore suspension, respectively.

5.3.2 基因克隆 Gene Clone

甘蔗抗坏血酸过氧化物酶基因 ScAPX1 的克隆和表达分析

通过克隆甘蔗 ScAPX1 及分析其在低温胁迫中的表达，为深入研究甘蔗 APX1 在低温胁迫中的功能及为甘蔗抗寒育种的分子机理提供依据（图15）。以甘蔗叶片总 RNA 为模板，通过 RT-PCR 技术克隆甘蔗叶片 ScAPX1 的完整 ORF 序列，采用生物信息学软件分析所克隆基因的编码蛋白特性，并用荧光定量PCR分析该基因在2个抗寒性差异较大的甘蔗品种GT28和YL6低温胁迫下的表达模式。结果表明，克隆得到甘蔗*ScAPX1*（NCBI登录号为KC794939），其包含一个759 bp的完整开放阅读框，编码252个氨基酸。该基因编码的蛋白不含信号肽，为可溶性蛋白，无跨膜结构，定位于细胞质，与高粱氨基酸的同源性为98%，推测其为胞质型抗坏血酸过氧化物酶基因。qRT-PCR分析结果表明，随着低温（0~4℃）时间的延长，2个甘蔗品种*ScAPX1*的表达量均是先上升后下降，但表达量存在差异，在整个胁迫过程中，抗寒强品种GT28的表达量始终比抗寒弱品种YL6高。甘蔗*ScAPX1*积极响应低温逆境胁迫，该基因的诱导表达与甘蔗品种本身的抗寒性密切相关。

（张保青，邵敏，黄玉新，黄杏，宋修鹏，陈虎，王盛，谭秦亮，杨丽涛*，李杨瑞*）

Cloning and expression analysis of peroxidase gene (*ScAPX1*) from sugarcane

Here cloning the *ScAPX1* of sugarcane and analyzing its expression under low temperature stress is aimed to provide a basis for further studies on the function of *APX1* gene in sugarcane under low temperature stress and investigating the molecular mechanism of breeding cold stress-resistant sugarcane (Fig. 15). Using total RNA of sugarcane leaf as template, RT-PCR technique was used to clone the complete ORF sequence of *ScAPX1* from sugarcane leaf, and analyze the characteristics of *ScAPX1* protein. Quantitative real-time PCR (qRT-PCR) method was applied to study the expressions of *ScAPX1* gene under low temperature stress in two sugarcane varieties GT28 and YL6 with widely different cold resistance. The results showed that the *ScAPX1* gene (GenBank accession number: KC794939) in sugarcane contained a complete open reading frame of 759 bp and encoded 252 amino acids. The protein encoded by this gene contained no signal peptide and no transmembrane structure, was a soluble protein and located in cytoplasm, and its amino acid homology with sorghum homologue was 98%. It was inferred that *ScAPX1* gene was a cytoplasmic ascorbic acid peroxidase gene (*ScAPX*). qRT-PCR analysis showed that with the extension of low temperature (0~4°C), the expression levels of *ScAPX1* gene in two sugarcane varieties increased first and then decreased, but their intensity of expression differed. In the whole process of cold stress, the relative expression level in GT28, a cold

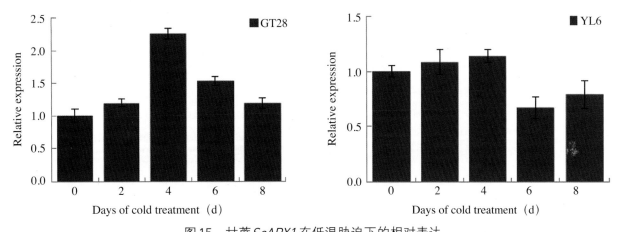

图15 甘蔗*ScAPX1*在低温胁迫下的相对表达

Fig. 15 The relative expression of *ScAPX1* under low temperature stress

resistant variety, was always higher than that in YL6, a cold sensitive variety. This result suggests that the *ScAPX1* gene of sugarcane is active in response to low temperature stress, and the induced expression of this gene is closely related to the cold resistance of sugarcane varieties.

(Bao-qing Zhang, Min Shao, Yu-xin Huang, Xing Huang, Xiu-peng Song, Hu Chen, Sheng Wang, Tan -liang Qin, Li-tao Yang[*], Yang-rui Li[*])

Molecular cloning and expression analysis of *ScTUA* gene in sugarcane

The aim of this study was to clone the full-length cDNA of sugarcane a-tubulin gene (*ScTUA*), investigate its sequence characteristics and analyze its expression in different resistant sugarcane varieties under cold stress. The *ScTUA* gene cDNA sequence was cloned from sugarcane leaf using RT-PCR techniques (Fig. 16). Its putative amino acid sequence was deciphered, and the expression of *ScTUA* gene in different resistant sugarcane varieties under cold stress was studied. The full-length cDNA of *ScTUA* (GenBank accession number: JQ230105) in sugarcane was cloned. The sequence included an open reading frame of 1 353 bp, encoding a polypeptide of 450 amino acids. Homology analysis showed that the deduced *ScTUA* protein was highly homologous to other *ScTUA* proteins from different species. Under cold stress, real-time PCR results showed that the expression tendency of

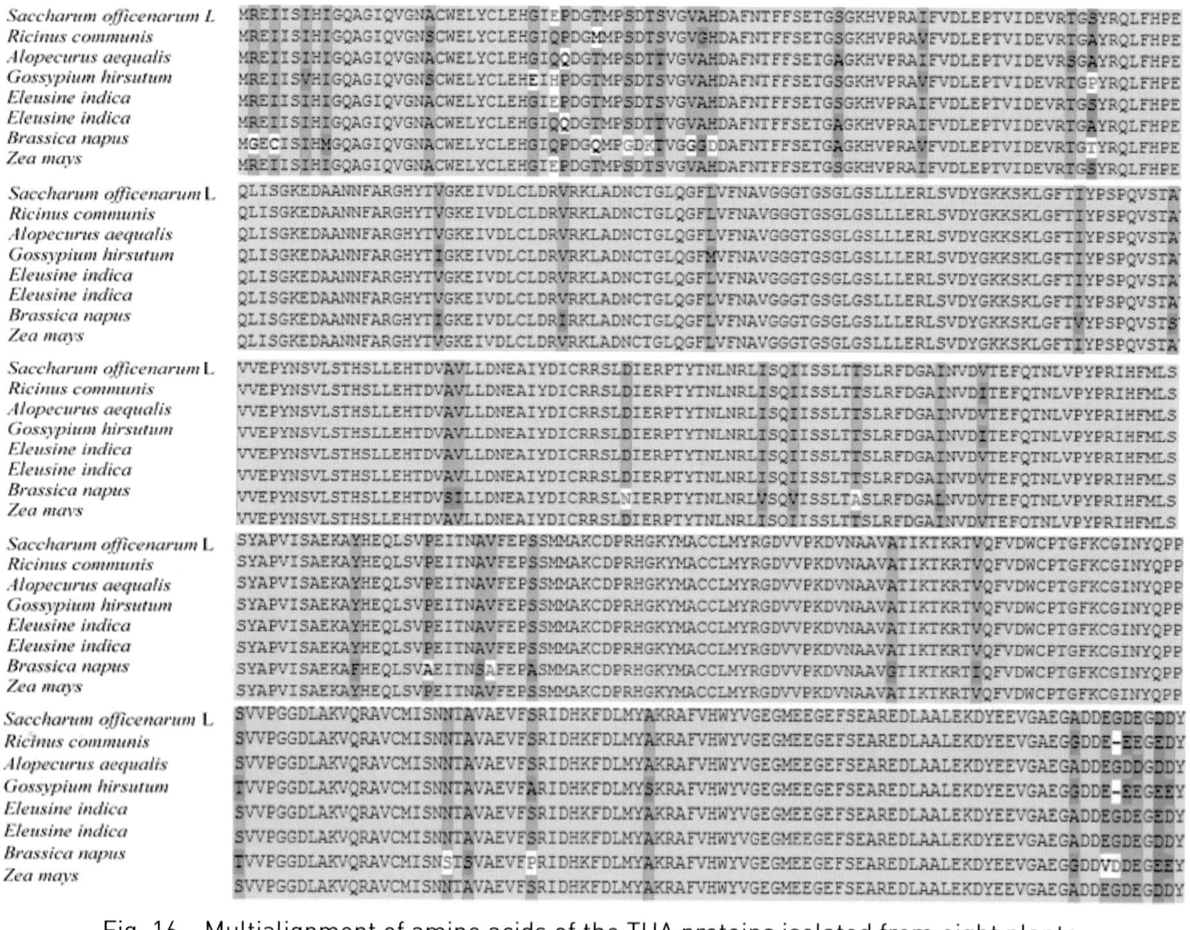

Fig. 16 Multialignment of amino acids of the TUA proteins isolated from eight plants

ScTUA gene in two sugarcane varieties was the same, but the relative expression level was clearly higher in cold-resistant variety GT28 than in cold-sensitive variety YL6. The activity of gene was altered in response to cold stress, but the functions of the gene were different, and the defense pathways involved were also different in different sugarcane varieties under cold stress, which provided an important molecular basis to strengthen sugarcane resistance to cold stress and other abiotic stress.

(Bao-qing Zhang, Min Shao, Yong-jian Liang, Xing Huang, Xiu-peng Song. Hu Chen. Li-tao Yang[*], Yang-rui Li[*])

用于克隆及分子标记分析的甘蔗高质量基因组DNA提取方法

甘蔗是世界上重要的糖料作物，其遗传背景复杂，基因组庞大复杂。但基因组测序尚未完成，严重制约了甘蔗分子生物学研究进展（图17）。为构建成熟完善的甘蔗高质量基因组DNA提取方法，本研究采用四种改良CTAB法提取甘蔗基因组DNA，先通过紫外分光光度计和琼脂糖凝胶电泳检测提取所得甘蔗基因组DNA的质量，再使用8对转座子保守区同源克隆扩增引物和8种基于单引物扩增反应的分子标记技术的78条单引物对提取的甘蔗基因组DNA的质量进行扩增验证。结果表明：（1）结合甘蔗基因组DNA的质量数据以及甘蔗转座子保守区同源克隆扩增和分子标记技术研究的验证结果来看，四种改良CTAB法的DNA提取效果表现依次为：方法三＞方法四＞方法二＞方法一。但方法三和方法四均使用到强腐蚀性的平衡酚而不安全环保，而方法二和方法一则分别使用了二次和一次氯仿抽提，根据样品的实际状态，本研究认为后两者中的一种可作为甘蔗基因组DNA的最佳提取方法；（2）建立了甘蔗6类转座子保守区同源克隆和8种基于单引物扩增反应的分子标记技术的扩增体系。本研究为以后开展甘蔗转座子的克隆鉴定利用和分子标记研究提供了帮助。

（刘俊仙，熊发前[*]，刘菁，罗丽，丘立杭，刘丽敏，吴建明，刘红坚，刘欣，卢曼曼，何毅波，李松[*]）

High quality sugarcane DNA extraction methods for cloning and molecular marker analysis

Sugarcane is an important sugar crop in the world. Its genetic background is complex and its genome is large and complex. However, due to incomplete genome sequencing, the progress of molecular biology in sugarcane has been restricted seriously (Fig. 17). In order to develop a method for extracting sugarcane genomic DNA with high quality, four improved CTAB methods were evalauted. The quality of the extracted sugarcane genomic DNA was first detected by UV spectrophotometer and agarose gel electrophoresis, respectively. Then, we used 8 pairs of transposon conserved region homolog cloning and amplification primers and 78 single primers of 8 molecule marker technologies based on single primer amplification reaction to amplify and verify the quality of the genomic DNA obtained. The results obtained are as follows: (1) According to the quality data of sugarcane genomic DNA, the ranking of DNA extraction methods were: Method three>Method four>Method two>Method one. However, the use of highly corrosive phenol in the methods three and four was not safe. The methods two and one used chloroform extraction twice and once respectively. We thus select method two or one as the optimum extraction method based on the sugarcane DNA sample quality and the safety of the method. (2) Using this DNA the amplification system of six types of homolog cloning and amplification of transposon conserved region and eight kinds of molecular marker technologies based on single primer amplification reaction in sugarcane has been established. This study would be helpful for the cloning, identification and utilization of sugarcane transposon and molecular marker research.

(Jun-xian Liu, Fa-qian Xiong[*], Jing Liu, Li Luo, Li-hang Qiu, Li-min Liu, Jian-ming Wu, Hong-jian Liu, Xin Liu, Man-man Lu, Yi-bo He, Song Li[*])

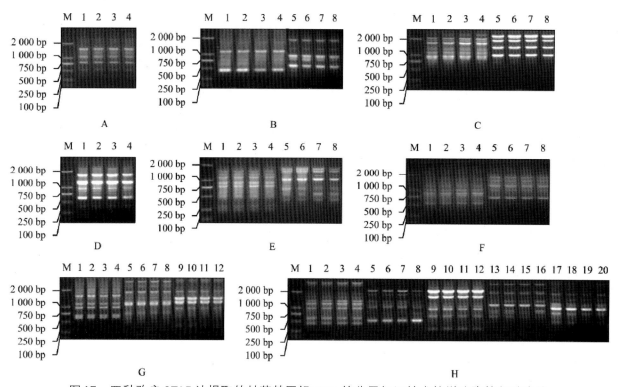

图17 四种改良CTAB法提取的甘蔗基因组DNA的分子标记技术扩增琼脂糖凝胶电泳

Fig. 17 The agarose gel electrophoresis of amplification products by molecular marker techniques of sugarcane genome DNA extracted by four improved CTAB methods

注：M: DL2000 DNA Marker; A: RAPD技术扩增的电泳，所用单引物为UBC001; B: ISSR技术扩增的电泳，所用单引物从左至右依次为8386和8082; C: BPS技术扩增的电泳，所用单引物从左至右依次为Bps1和Bps2; D: DAMD技术扩增的电泳，所用单引物为6.2H (-); E: URP技术扩增的电泳，所用单引物从左至右依次为URP9F和URP25F; F: SCoT技术扩增的电泳，所用单引物从左至右依次为SCoT12和SCoT13; G: iPBS技术扩增的电泳，所用单引物从左至右依次为2095、2217和2218; H: CBDP技术扩增的电泳，所用单引物从左至右依次为CAAT1、CAAT2、CAAT3、CAAT4和CAAT18

Note: M: DL2000 DNA Marker; A: Electrophoretic analysis of RAPD amplification, the single primer used was UBC001; B: Electrophoretic analysis of ISSR amplification, the single primers used from left to right were 8386 and 8082 respectively; C: Electrophoretic analysis of BPS amplification, the single primers used in turn from left to right were Bps1 and Bps2 respectively; D: Electrophoretic analysis of DAMD amplification, the single primer used was 6.2H(-); E: Electrophoretic analysis of SCoT amplification, the single primers used were SCoT12 and SCoT13 respectively; F: Electrophoretic analysis of URP amplification, the single primers used from left to right were URP9F and URP25F respectively; G: Electrophoretic analysis of iPBS amplification, the single primers used in turn from left to right were 2095, 2217 and 2218 respectively; H: Electrophoretic analysis of CBDP amplification, the single primers used in turn from left to right were CAAT1, CAAT2, CAAT3, CAAT4, and CAAT18, respectively

甘蔗乙烯信号转导途径关键基因*CTR1*的克隆与表达分析

为了研究*CTR1*基因与甘蔗糖分积累的关系，本研究利用同源克隆的方法，以甘蔗品种

GT28幼嫩叶片的cDNA为模板，成功克隆得到2 103 bp的甘蔗*CTR1*基因（*ScCTR1*; GenBank注册号为MK158246）。*ScCTR1*基因的开放阅读框（ORF）序列长1 656 bp，编码551个氨基酸，预测蛋白质的分子质量为62.78 kD（图18）。*ScCTR1*蛋白具有*CTR1*同源蛋白典型的丝氨酸/苏氨酸蛋白激酶结构域。实时荧光定量PCR结果表明，*ScCTR1*基因在茎和叶中都有表达，在茎中其表达量呈先增加后降低的趋势，在未成熟叶、成熟叶和老叶中，其基因表达呈降低的趋势。本研究结果对阐明乙烯调控甘蔗茎糖分积累和叶发育的机理提供了重要的参考依据。

（王凡伟，肖冬，李杨瑞*，何龙飞*，王爱勤*）

Molecular cloning and expressional analysis of the gene of Key Factor *CTR1* in sugarcane ethylene signal transduction pathway

In order to study the relationship between *CTR1* gene and sugar accumulation in sugarcane, in the present study, we successfully cloned a 2 103 bp of sugarcane *CTR1* gene (*ScCTR1*; GenBank registration number is MK158246) from tender leaves of a sugarcane cultivated varieties GT28 by homologous cloning method (Fig. 18). The open reading frame sequence of *ScCTR1* is 1 656 bp in length, encoding 551 amino acids. The molecular mass of deduced sugarcane *CTR1* protein is 138.3 kD, and the pI is 5.90. Protein domain analysis showed that *ScCTR1* had a serine/threonine protein kinase domain which is a typical domain in *CTR1* homologous proteins. qRT-PCR results showed that *ScCTR1* was expressed in both stem and leaf, and the expression level of *ScCTR1* in stem increased first and then

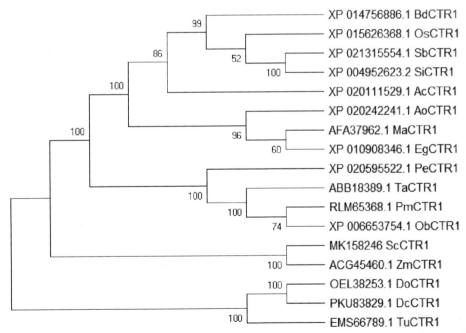

图18 甘蔗*ScCTR1*蛋白的系统进化树

Fig. 18 Phylogenetic tree of *ScCTR1* protein from sugarcane

注：Bd: 二穗短柄草；Os: 水稻；Sb: 高粱；Si: 小米；Ac: 菠萝；Ao: 芦笋；Ma: 小果野芭蕉；Eg: 油棕；Pe: 桃红蝶兰；Ta: 小麦；Pm: 黍；Ob: 短花药野生稻；Sc: 甘蔗；Zm: 玉米；Do: Ⅱ型少花古尔德草；Dc: 铁皮石斛；Tu: 图小麦

Note: Bd: *Brachypodium distachyon*; Os: *Oryza sativa*; Sb: *Sorghum bicolor*; Si: *Setaria italica*; Ac: *Ananas comosus*; Ao: *Asparagus officinalis*; Ma: *Musa acuminata*; Eg: *Elaeis guineensis*; Pe: *Phalaenopsis equestris*; Ta: *Triticum aestivum*; Pm: *Panicum miliaceum*; Ob: *Oryza brachyantha*; Sc: *Saccharum hybrid cultivar*; Zm: *Zea mays*; Do: *Dichanthelium oligosanthes*; Dc: *Dendrobium catenatum*; Tu: *Triticum urartu*

decreased, while in immature leaves, maturing leaves and matured leaves, the gene expression showed a decreasing trend. The result from this study provides an important reference to illustrate the mechanism of ethylene regulated sucrose accumulation in stalks and leaf development of sugarcane.

(Fan-wei Wang, Dong Xiao, Yang-rui Li[*], Long-fei He[*], Ai-qin Wang[*])

Functional analysis of *Leifsonia xyli* subsp. *xyli* membrane protein gene *Lxx18460* (anti-sigma K)

Sugarcane is an important sugar and economic crop in the world. Ratoon stunting Disease (RSD) of sugarcane, caused by *Leifsonia xyli* subsp. *xyli*, is widespread in countries and regions where sugarcane is grown and also limited to sugarcane productivity (Fig. 19). Although the whole genome sequencing of *Leifsonia xyli* subsp. *xyli* was completed, progress in understanding the molecular mechanism of the disease has been slow because it is difficult to grow in culture. The *Leifsonia xyli* subsp. *xyli* membrane protein gene *Lxx18460* (anti-sigma K) was cloned from the *Lxx*-infected sugarcane cultivar GT11 at the mature stage using RT-PCR technique, and the gene structure and expression in infected sugarcane were analyzed. The *Lxx18460* gene was transformed into *Nicotiana tabacum* by *Agrobacterium tumefaciens*-mediation. The transgenic tobacco plants overexpressing *Lxx18460* had lower levels in plant height, leaf area, net photosynthetic rate and endogenous hormones of IAA, ABA and GA3, as well as lower activities of three antioxidant enzymes, superoxide dismutase (SOD), peroxidase (POD) and catalase (CAT) than the wild type (WT) tobacco. With the plant growth, the expression of *Lxx18460* gene and protein was increased. To better understand the regulation of *Lxx18460* expression, transcriptome analysis of leaves from transgenic and wild type tobacco was performed. A total of 60 222 all-unigenes were obtained through BGISEQ-500 sequencing. Compared the transgenic plants with the WT plants, 11 696 upregulated and 5949 downregulated genes were identified. These differentially expressed genes involved in many metabolic pathways including signal transduction, biosynthesis of other secondary metabolism, carbohydrate metabolism and so on. Though the data presented here are from a heterologous system, *Lxx 18460* has an adverse impact on the growth of tobacco; it reduces the photosynthesis of tobacco, destroys the activity of defense enzymes, and affects the levels of endogenous hormones, which indicate that *Lxx18460* may act important roles in the course of infection in sugarcane. Conclusions:

Fig. 19 Phenotypes of *Lxx18460* transgenic and WT tobacco plants. a, b and c *Lxx18460* transgenic and WT plants at 50, 80 and 110 days after emergence, respectively

This is the first study on analyzing the function of the membrane protein gene *Lxx18460* of antisigma K (σK) factor in *Leifsonia xyli* subsp. *xyli*. Our findings will improve the understanding of the interaction between the RSD pathogen *Leifsonia xyli* subsp. *xyli* and sugarcane. The output of this study will also be helpful to explore the pathogenesis of RSD.

(Kai Zhu, Min Shao, Dan Zhou, Yong-xiu Xing, Li-tao Yang[*], Yang-rui Li[*])

5.3.3 转基因研究 Genetic Modification

低温胁迫对转 SoTUA 基因甘蔗生理生化特性的影响

研究低温胁迫对转α-微管蛋白（TUA）基因甘蔗生理生化特性的影响，对甘蔗TUA基因（*SoTUA*）进行功能鉴定，为甘蔗抗寒品种选育提供参考依据（图20）。通过对pCAMBIA3300载体上的标记基因*GFP*进行PCR扩增及测序，鉴定转*SoTUA*基因阳性甘蔗品系。从中选择4个转基因品系（L1、L2、L3和L4），以野生型ROC22植株为对照（CK），对各处理进行低温（4℃）胁迫处理，在胁迫第0、4、7和12 d时分别测定不同处理甘蔗的生理生化指标。运用PCR筛选得到转*SoTUA*基因T1代甘蔗植株，对其中4个品系及CK的低温胁迫试验结果表明，在低温下各品系可溶性糖含量以不同模式变化，转基因各品系普遍高于CK；各品系中的可溶性蛋白在冷驯化前期强烈诱导，总体变化趋势为先升高后降低，转基因各品系普遍高于CK；各品系中丙二醛（MDA）含量增加，其中CK的增幅最大；过氧化物酶（POD）活性下降，除L2外，POD活性变化总体上呈先升高后降低的变化趋势；脯氨酸含量在低温胁迫过程中的变化趋势无明显规律。利用改进的极点排序法对各品系抗寒性进行评价，可知各品系抗寒性强弱排序为：L3>L1>L2>L4>CK。低温胁迫下4个转*SoTUA*基因甘蔗品系在生理生化指标上的表现均优于CK，即*SoTUA*基因在甘蔗体内的过量表达有利于甘蔗抗寒性能的提高。

（唐丽华，黄婵，陈教云，杨丽涛，邢永秀，农友业，李杨瑞[*]）

Effects of low temperature stress on physiological and biochemical characteristics of *SoTUA* transgenic sugarcane

This study was conducted to investigate the effects of low temperature stress on physiological and biochemical characteristics of α-tubulin (TUA) transgenic sugarcane, and to identify the function of sugarcane TUA gene (*SoTUA*). It provided reference for sugarcane cold-resistant variety breeding (Fig. 20). *SoTUA* transgenic sugarcane lines were identified by PCR amplification and sequencing of GFP marker gene on vector pCAMBIA3300. Four transgenic lines (L1, L2, L3 and L4) were chosen for a low temperature (4 °C) stress assay when the wild type of sugarcane variety ROC22 was used as control (CK). The changes of physiological and biochemical indicesindices were measured at 0, 4, 7, 12 d after treatment. The *SoTUA* transgenic T1 sugarcane plants were screened by PCR, and the results of low temperature stress assay on four lines and the control showed that the soluble sugar content of each line varied in different patterns at low temperature, and the transgenic lines were generally higher than the CK. Soluble protein of all lines was strongly induced in the early stage of cold treatment, the overall change trend was first increased and then decreased, and the transgenic lines were generally higher than the CK. Malonaldehyde (MDA) content in each line increased, especially in the control. Peroxidase

(POD) activity decreased, except for L2, POD activity generally increased and then decreased. The proline content change had no obvious regulations during the low temperature stress. The improved pole ordination method was used for the comprehensive evaluation of cold resistance. The results showed that the cold resistance of all the lines was ranked as L3>L1>L2>L4>CK. The physiological and biochemical performances of the four *SoTUA* transgenic sugarcane lines were better than the CK under low temperature stress. Therefore, it can be concludedthat the overexpression of *SoTUA* gene in sugarcane is conducive to the improvement of cold resistance.

(Li-hua Tang, Chan Huang, Jiao-yun Chen, Li-tao Yang, Yong-xiu Xing, You-ye Nong, Yang-rui Li[*])

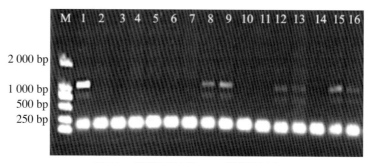

图20 转基因甘蔗PCR检测结果

Fig.20　PCR detection of transgenic sugarcane

注：M. DL2000 DNA Marker；1. pCAMBIA3300-SoTUA-GFP 重组质粒（阳性对照）；2. CK（阴性对照）；3. H_2O；4~16. 转基因甘蔗品系（其中8、9、12和13分别对应株系L1、L2、L3和L4）

Note: M. DL2000 DNA Marker; 1. Recombinant plasmid pCAMBIA3300-So-TUA-GFP (positive control); 2. CK (negative control); 3. H_2O; 4-16. Transgenic sugarcane lines (in which 8, 9, 12 and 13 were corresponding to L1, L2, L3 and L4, respectively)

5.4　甘蔗病虫害致病机理及生物防治 Pathogenesis Mechanism and Biological Control of Sugarcane Diseases and Pests

5.4.1　甘蔗虫害生物防治 Biological Control of Sugarcane Pests

The Impact of *Cryptolaemus montrouzieri* Mulsant (Coleoptera: Coccinellidae) on control of *Dysmicoccus neobrevipes* Beardsley (Hemiptera: Pseudococcidae)

Cryptolaemus montrouzieri (Coleoptera: Coccinellidae) is an important predator of the mealybug *Dysmicoccus neobrevipes* (Hemiptera: Pseudococcidae), a major pest of *Agave sisalana* in China. Limited reports on the efficacy of *C. montrouzieri* against *D. neobrevipes* are available. This study reports the predatory ecacy and functional response of *C. montrouzieri* against *D. neobrevipes* under laboratory conditions. The prey consumption rate per day of 4th instar larvae of *C. montrouzieri* feeding on 1st instar *D. neobrevipes* nymphs (241.3 mealybugs) was the highest among the different larval life stages of the beetle. For *C. montrouzieri*, the prey consumption per day of adult females (19.8 mealybugs) was significantly higher compared to males (15.2 mealybugs) when feeding on 3rd instar *D.*

neobrevipes nymphs. The functional responses of *C. montrouzieri* on 1st and 2nd instar *D. neobrevipes* nymphs were determined as Holling type II. The search rates of *C. montrouzieri* 4th instar larvae towards the 1st and 2nd instar nymphs of *D. neobrevipes* were higher than those of the other beetle life stages. In addition, the handling times of 4th instar larvae were shorter than those of the other beetle life stages. The results from this study indicate that *C. montrouzieri* can be used as a predator of *D. neobrevipes* and, therefore, it should be evaluated further for use as a biocontrol agent in *D. neobrevipes* management programs.

(Zhen-qiang Qin, Jian-hui Wu, Bao-li Qiu, Shaukat Ali, Andrew G. S. Cuthbertson)

2015—2017年广西桂林柑橘黑刺粉虱种群动态

本研究以广西桂林柑橘园黑刺粉虱（*Aleurocanthus spiniferus*）为对象，采用田间调查该粉虱卵、若虫、蛹和成虫的方法，研究了该粉虱的种群消长规律（图21）。结果表明：广西桂林柑橘园每年3月中旬至12月上旬的气温可满足黑刺粉虱的生长发育需要，2015—2017年间黑刺粉虱种群变动较大，黑刺粉虱卵至蛹全年发生高峰在5月中旬至7月上旬，其次在8月下旬至9月上旬；而成虫全年发生高峰在5月上旬，其次则在9月上中旬。本研究表明每年5～9月是桂林柑橘园黑刺粉虱自然种群增长较快的时期，生产上应用密切关注这一时期的该粉虱种群动态，并根据其危害情况及时采取必要的防治措施。

（覃振强，张戈壁，吴建辉，邱宝利）

Population dynamic of *Aleurocanthus spiniferus* (Quaintance) at citrus orchard in Guilin, Guangxi from 2015 to 2017.

The population dynamic of *Aleurocanthus spiniferus* (Quaintance) (Hemiptera: Aleyrodidae) was analyzed according to the investigation of eggs, nymphs, pupa and adults at citrus orchards in Guilin, Guangxi Province in this study (Fig. 21). The results showed that temperature from mid-March to early December at citrus orchards in Guilin could meet the requirement of the growth and development of *A. spiniferus*. The population of *A. spiniferus* had great changes during 2015 to 2017. The peak population from eggs to pupa of A. spiniferus was from mid-May to early July, followed by late August to early

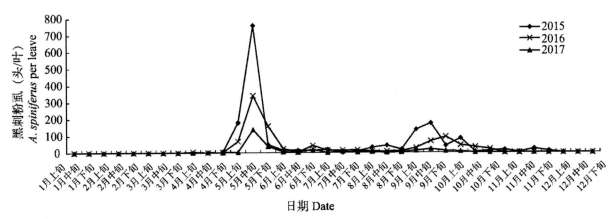

图21 2015—2017年广西桂林柑橘黑刺粉虱成虫种群动态

Fig.21 Population dynamics of A.spiniferus adults at citrus orchard in Guilin, Guangxi from 2015 to 2017

September. While the peak population of A. spiniferus adults was in early May, and followed by early to mid-September. The season of higher population of A. spiniferus was from May to September, when prevention and controlling measures should be taken according to the population dynamic of the pest.

(Zhen-qiang Qin, Ge-bi Zhang, Jian-hui Wu, Bao-li Qiu)

温度对甘蔗条螟生长发育和繁殖的影响

为明确温度对甘蔗条螟 Chilo sacchariphagus 生长发育和繁殖的影响，在20℃、23℃、26℃、29℃和32℃不同温度下测定甘蔗条螟各虫态的发育历期和成虫繁殖力（表37）。结果显示，在20～32℃范围内，各虫态发育历期随温度升高而缩短，20℃下完成1个世代需要83.30d，而32℃下仅需35.56d。各虫态的发育速率与温度呈显著正相关，符合Logistic回归模型。卵期、幼虫期、蛹期、成虫产卵前期以及全世代的发育起点温度分别为15.87℃、9.90℃、13.67℃、10.79℃和11.27℃，有效积温分别为69.49d·℃、490.99d·℃、142.95d·℃、26.72d·℃和717.68d·℃。雌、雄成虫寿命均随温度的升高逐渐缩短，雌成虫产卵量在26℃时最高为136.37粒/头。卵和蛹在26℃时的存活率最高，分别为94.71%和93.55%，在29℃时幼虫的存活率最高，为47.73%，在26℃时全世代的存活率最高，为42.08%。表明温度是影响甘蔗条螟生长发育和繁殖的关键因素，26～29℃是甘蔗条螟生长发育和繁殖的适宜温度。

（魏吉利，潘雪红[*]，黄诚华，商显坤，林善海）

Effects of temperature on the development and reproduction of spotted borer Chilo sacchariphagus (Lepidoptera: Pyralidae).

In order to determine the effects of temperature on the development and reproduction of spotted borer Chilo sacchariphagus, the durations of different developmental stages and reproductive capacity of C. sacchariphagus were measured at 20°C, 23°C, 26°C, 29°C and 32°C(Table 37). The results showed that the average durations of various developmental stages of C. sacchariphagus shortened along with increasing temperature from 20 to 32°C. The durations to complete one lifecycle were 83.30 d at 20°C, and 35.56 d at 32°C for C. sacchariphagus. The developmental rate of each developmental state was positively correlated with the temperature, which was consistent with the Logistic regression model. The developmental threshold temperatures for egg stage, larval stage, pupal stage, preoviposition stage and the whole generation were 15.87°C, 9.90°C, 13.67°C, 10.79°C and 11.27°C, respectively. The effective accumulated temperatures were 69.49d·°C, 490.99d·°C, 142.95d·°C, 26.72 d·°C and 717.68d·°C, respectively. The adult longevity shortened with increasing temperature, and the highest oviposition per moth (136.37 eggs) was observed at 26°C. The highest survival rates of eggs and pupae at 26°C were 94.71% and 93.55%, respectively, and the highest survival rate of the larva was 47.73% at 29°C, and that of the whole generation was 42.08 % at 26°C. The results indicated that temperature was the key factor affecting the development and reproduction of C. sacchariphagus, and the optimum temperature for development and reproduction ranged from 26°C to 29°C.

(Ji-li Wei, Xue-hong Pan[*], Cheng-hua Huang, Xian-kun Shang, Shan-hai Lin)

表37 不同温度条件下甘蔗条螟各虫态及全世代发育历期

Table 37 Developmental durations of different stages and generations of
Chilo sacchariphagus at different temperatures

发育阶段 Developmental stage	20℃	23℃	26℃	29℃	32℃
卵期 Egg	15.32±0.08a	11.36±0.07b	6.21±0.05c	5.12±0.03d	4.55±0.02e
幼虫期 Larva	48.39±0.46a	36.42±0.69b	30.85±1.22c	26.61±0.63d	21.79±0.33e
蛹期 Pupa	18.39±0.19a	13.05±0.16b	10.35±0.09c	9.18±0.22d	8.03±0.17e
成虫产卵前期 Preoviposition stage	3.25±0.37a	2.00±0.15b	1.60±0.14b	1.45±0.14b	1.38±0.13b
全世代 Whole generation	83.30±0.94a	61.81±0.93b	47.86±0.56c	39.33±0.62d	35.56±0.55e

注：表中数据为平均数±标准误。同行数据后不同字母表示经Duncan新复极差法检验在$P<0.05$水平差异显著。

Note: Data are mean±SE. Different letters in the same row indicate significant difference at $P<0.05$ level by Duncan's new multiple range test.

Soil pH and organic matter levels in the area of occurrence of the whitegrub *Alissonotum impressicolle* Arrow (Coleoptera: Dynastidae).

The whitegrub *Alissonotum impressicolle* Arrow is an important underground pest of sugarcane in Guangxi, China. Soil physicochemical properties were analysed to determine suitable areas for the occurrence of *A. impressicolle*. Soils 5-20 cm deep were collected from different ecological sugarcane areas and analysed for pH and organic matter content. In the north of Guangxi soils were acidic, with the pH at some sites less than 5.5. Most soils in the central sugarcane growing area were neutral pH. There were acid, neutral and alkaline soils in the western and southwest of Guangxi, but most of them were pH neutral. Most soils were rich in organic matter, 77.55% of them with over 20 g/kg. No significant difference in pH values was detected between paddy fields and sloping fields, but the organic matter content of paddy fields was higher than that of sloping fields. Among the 49 soil samples, 30.61% and 34.69%, were slightly acidic soils and neutral soils, respectively, and 65.30% had organic matter in the range of 20 g/kg to 40 g/kg. Our results indicated that neutral or slightly acidic soils with abundant organic matter were more attractive to *A. impressicolle*. Although there were different types of soil pH and organic matter in different ecological sugarcane growing areas in Guangxi, *A. impressicolle* has strong adaptability to soil pH and organic matter and could survive in all sugarcane planting areas in Guangxi.

(Xian-kun Shang, Cheng-hua Huang[*], Ji-li Wei, Xue-hong Pan, Shan-hai Lin)

温度对蔗螟天敌等腹黑卵蜂生长发育的影响

探明温度对广西蔗区甘蔗螟虫优势寄生性天敌等腹黑卵蜂生长发育的影响，为该天敌的室内人工饲养繁殖、田间保护及其对蔗螟的有效防控提供理论基础（表38）。测定了等腹黑卵蜂在17℃、20℃、23℃、26℃、29℃和32℃ 6个恒温条件下的世代发育历期和发育速率，利用线性回归、S曲线回归和Logistic回归分析了温度与等腹黑卵蜂发育速率之间的关系，并用直接回归法和直接最优法2种方法比较了等腹黑卵蜂的发育起点温度和有效积温。等腹黑卵蜂在17～32℃范围内均能正常发育，发育历期随温度的升高而缩短，发育速率随温度的升高而加快。Logistic模型更能拟合等腹黑卵蜂发育速率与温度的关系，其回归方程为 V= 0.175 5/（1+ e4.016 6-0.152 9T）。经 Logistic 模型计算出，等腹黑卵蜂的发育最适温度为 26.27℃相应发育历期为 11.40d。利用直接最优法计算得出等腹黑卵蜂发育起点温度和有效积温分别为 11.25℃ 和 172.31 d·℃。温度是影响等腹黑卵蜂生长发育的重要因素，本研究结果为今后等腹黑卵蜂的保护和甘蔗螟虫的天敌防治提供了科学依据。

（潘雪红，商显坤，魏吉利，黄诚华[*]）

Effect of Temperature on Development of Natural Enemy *Telenomus dingus*

Telenomus dingus is a kind of dominant natural enemy, which plays an important role to stem borer control in Guangxi. In order to provide a foundation for artificial rearing of *T. dingus* in the laboratory and protecting the natural enemy in the fields to control sugarcane borers, the effects of temperature on the development of *T. dingus* were investigated (Table 38). The developmental duration and developmental rate of *T. dingus* under different temperature condition (17°C, 20°C, 23°C, 26°C, 29°C and 32°C) were studied. The relationships between the developmental rate and temperature were also analyzed by using the three methods of Linear, S and Logistic regression. And the developmental threshold temperature and effective accumulated temperature were made out and compared by using the linear regression method and the direct optimal method. *T. dingus* could development normally under the temperature from 17 to 32°C, the developmental duration decreased under high temperature conditions, and the developmental rate accelerated with the increasing temperature. The Logistic model could fit the relationships between developmental rate and temperature better, the regression equation was V=0.175 5/(1+e4.016 6-0.152 9T) . The optimal temperature was 26.27°C and developmental duration was 11.40 d. The analysis by direct optimal method, the developmental threshold temperature and the effective accumulated temperature was 11.25°Cand 172.31 d·°C respectively. Temperature was very important in the development of *T. dingus*, and the optimum temperature range for its development. These findings will provide a scientific basis for rearing in the laboratory and protecting the natural enemy in the fields to control sugarcane borers.

（Xue-hong Pan, Xian-kun Shang, Ji-li Wei, Cheng-hua Huang[*]）

表38　不同温度下等腹黑卵蜂的发育历期和发育速率
Table 38　Developmental duration and developmental rate of *T. dignus* at different temperatures

温度(℃) Temperature	平均历期 N (d) Average duration	发育速率 V(1/d) Developmental rate
17	27.96±0.21a	0.035 8
20	20.54±0.88b	0.048 7
23	16.20±0.39c	0.061 7
26	11.67±0.20d	0.085 7
29	9.17±0.17e	0.109 1
32	8.10±0.10e	0.123 5

5.4.2 甘蔗病害防治及机理研究
Sugarcane Disease Control and Mechanism Research

甘蔗黑穗病及其防治研究进展

甘蔗是世界上重要的糖料和生物质能源作物，由甘蔗鞭黑粉菌（*Sporisorium scitamineum*）引起的甘蔗黑穗病是甘蔗生产上的主要病害，严重影响甘蔗产量和质量，对我国乃至世界甘蔗生产和蔗糖产业的可持续性发展构成了严重威胁（表39）。在分子水平上甘蔗鞭黑粉菌致病机理和甘蔗抗病机理研究已成为当前研究的热点，深入的机理研究为甘蔗黑穗病的有效防治和抗病育种提供科学依据。目前，选育和推广抗黑穗病新品种是防治甘蔗黑穗病最经济有效的措施，而我国在抗病育种的研究基础比较薄弱。综述了甘蔗黑穗病发生与危害、病原特征、遗传多样性、致病机理、抗病育种及防治对策，并对甘蔗黑穗病研究中存在的问题与今后的研究思路进行了探讨与展望。

（韦金菊，宋修鹏，魏春燕，张小秋，黄伟华，颜梅新[*]）

Research progress on sugarcane smut and its control

Sugarcane is one of the important sugar and biomass energy crops in the world. Sugarcane smut caused by *Sporisorium scitamineum* is the main disease of sugarcane, which results in enormous yield losses and decline in cane quality, and poses a serious threat to the sustainable development of sugarcane production and sugar industry in China and even in the world (Table 39). Study on pathogenesis of *S. scitamineum* and resistance mechanism of sugarcane at molecular level is expected to provide a scientific basis for smut control and disease resistance breeding of sugarcane. At present, breeding and promotion of new resistant varieties is the most economical and effective measure to control sugarcane smut. However, research on breeding resistance cultivars is relatively small in China. This paper reviewed disease occurrence, pathogenic feature, genetic diversity, pathogenic mechanism, disease resistance breeding and control measures. The research ideas of sugarcane smut were also discussed according to current research.

(Jin-ju Wei, Xiu-peng Song, Chun-yan Wei, Xiao-qiu Zhang, Wei-hua Huang, Mei-xin Yan[*])

表39 甘蔗鞭黑粉菌多样性

Table 39 Genetic diversity of *Sporiscrium scitamineum*

病菌来源地区 Source area of sugarcane smut	分析方法 Analysis methods	分析结果 Analysis results	作者 Authors
南非、夏威夷、留尼汪岛	RAPD、ITS	主要带型无差异，遗传多样性水平极低	Sing等，2005
亚洲、非洲、澳大利亚、美国等13个国家或地区	AFLP	总体遗传水平较低，但是来自于东南亚菲律宾、泰国及中国台湾地区的甘蔗鞭黑粉菌遗传分化较丰富	Braithwaite等，2004
亚洲、非洲、美洲等15个国家	微卫星 (Microsatellities)	美洲和非洲甘蔗鞭黑粉菌群体的遗传多样性水平极低，而亚洲群体遗传多样性水平较高	Raboi等，2007

(续)

病菌来源地区 Source area of sugarcane smut	分析方法 Analysis methods	分析结果 Analysis results	作者 Authors
福建、云南、广东、广西、海南、江西	RAPD	遗传分化较丰富，与地理来源具有一定相关性，具地域性特点	阙友雄等，2004
中国华南蔗区	ISSR、RAPD	分子遗传多样性水平属中等，从分子水平揭示华南蔗区可能可能有新的生理小种	Shen等，2012
广西（南宁、武宣、宜州、隆安、扶绥、贵港、龙州、横县、柳州）	RAPD	广西甘蔗鞭黑粉菌遗传分化度较低	林珊宇等，2016

广西甘蔗白条病病原菌的分离鉴定

为明确引起我国甘蔗检疫性细菌病害甘蔗白条病的病原菌及其症状特点，从广西来宾和崇左甘蔗主产区采集疑似白条病病害样本，采用组织分离法进行病原菌的分离，利用形态学与基因序列分析相结合的方法鉴定病原菌，柯赫氏法则回接验证病原菌的致病性（图22）。结果表明，引起广西甘蔗白条病的病原菌为白条黄单胞杆菌（*Xanthomonas albilineans*），16S-23S rRNA基因转录间隔区（ITS）基因序列分析以及白条黄单胞杆菌特异基因分析结果显示分离到的白条病病原菌与法国、巴西以及新西兰等国家的白条病病原菌同源性达到99%～100%。人工接种条件下，该病原菌还可侵染玉米并发生类似病症。本研究成功分离鉴定我国广西甘蔗白条病病原菌，为今后的致病机理研究和白条病抗病育种等方面奠定基础。

（魏春燕，韦金菊，张小秋，张保青，宋修鹏，李德伟，覃振强*，李杨瑞）

Isolation and identification of the pathogens causing sugarcane leaf scald in Guangxi Province, China.

In order to clarify the symptoms and pathogens of the bacterial quarantine disease, sugarcane leaf scald disease, infected samples showing suspected symptoms were collected from Laibin and Chongzuo, main sugarcane growing areas in Guangxi Province. Pathogenic isolates were obtained and purified by tissue isolation method, identified based on morphological characteristics and gene sequencing analysis (Fig. 22). Their pathogenicity was proved by the Koch's. The results showed that the causal agent of sugarcane leaf scald disease in Guangxi is Xanthomonas albilineans. 16S-23S ribosomal RNA intergenic spacer (ITS) and specific gene sequence analysis results showed that leaf scald pathogen isolated from Guangxi shared 99% ~ 100% similarity with the pathogen isolated in France, Brazil and New Zealand. Under artificial inoculation conditions, the pathogen can also infect corn and develop similar disease symptoms. This study isolated and identified the pathogen of sugarcane leaf scald in Guangxi, China. The results lay a foundation for future research on disease mechanism of leaf scald and disease resistance breeding.

(Chun-yan Wei, Jin-ju Wei, Xiao-qiu Zhang, Bao-qing Zhang, Xiu-peng Song, De-wei Li, Zhen-qiang Qin*, Yang-rui Li)

图22 甘蔗白条病的甘蔗病株的田间症状和分离菌株形态特征

Fig. 22 Symptom of leaf scald infected sugarcane in field and the morphology of the isolated pathogen strain

注：A～C. 甘蔗白条病的甘蔗病株的田间叶部症状（品种：桂糖46号）；D～F. 甘蔗白条病的甘蔗病株的田间茎部症状（品种：桂糖46号）；G, H. 分离菌株在XAS平板上的培养性状（6 d）。

Note: A~C show the leaf symptoms of sugarcane variety GT46 infected by leaf scald in the field. D～F show the stem symptoms of sugarcane variety GT46 infected by leaf scald in the field. G and H show the morphology of the isolated strains cultured on XAS medium for 6 days.

桂糖甘蔗新品系黑穗病抗性鉴定

防治甘蔗黑穗病最有效的途径是种植抗病品种，而甘蔗抗黑穗病评价则是抗病品种选育过程中重要程序（表40）。本研究对广西农科院甘蔗研究所选育的8个桂糖甘蔗新品系进行人工浸渍接种甘蔗鞭黑粉菌混合冬孢子悬浮液，同时调查大田自然发病情况，收集一新一宿黑穗病发病情况，以新台糖22号（ROC22）为对照品种，综合评价桂糖甘蔗新品系对黑穗病的抗性。综合新植、宿根的人工接种和大田自然发病评价结果，抗性类型为抗病的品系有2个，分别为桂糖12-765和桂糖12-2262；抗性类型为中抗的品系有2个，分别为桂糖12-2476和桂糖12-2004；抗性类型为中感的品系有1个，为桂糖12-162；抗性类型为感病的品系3个，分别为桂糖12-762、桂糖12-2425和桂糖12-917；对照品种ROC22的综合抗性类型为感病。对比人工接种和自然发病结果，人工接种黑穗病能更准确评价甘蔗品种抗黑穗病的水平，为选育高产高糖高抗黑穗病甘蔗品种提供依据。

（经艳，周会，刘昔辉，谭芳，张小秋，张荣华，宋修鹏，李杨瑞，颜梅新，雷敬超，李海碧，覃振强，罗亚伟，李冬梅，韦金菊[*]）

Smut resistant identification in new Guitang sugarcane clones

The most effective way to control sugarcane smut is to plant resistant varieties, and the evaluation of resistance to sugarcane smut is a very important procedure in the breeding of resistant varieties (Table 40). In order to evaluate the resistance to smut of the eight new sugarcane clones bred by Sugarcane Research Institute of Guangxi Academy of Agricultural Sciences, the suspension of mixed *Sporisorium scitamineum* were inoculated to the new clones to obtain the incidence, also the natural incidences of smut in the field were investigated. The Investigation period included new planting and first ratoon of sugarcane. ROC22 was used as control. The results showed that GT12-765 and GT12-2262 was resistant to smut; the clones GT12-2476 and GT12-2004 were moderately resistance; GT12-162 was moderately susceptibility; GT12-762, GT12-2425 and GT12-917 were susceptible to smut; and the CK (ROC22) was susceptible. The artificial inoculation can more accurately evaluate the resistance to sugarcane smut, compared to the natural infection. The results of this study provide a basis for breeding with high resistance to smut of sugarcane varieties.

(Yan Jing, Hui Zhou, Xi-hui Liu, Fang Tan, Xiao-qiu Zhang, Rong-hua Zhang, Xiu-peng Song, Yang-rui Li, Mei-xin Yan, Jing-chao Lei, Hai-bi Li, Zhen-qiang Qin, Ya-wei Luo, Dong-mei Li, Jin-ju Wei*)

表40 甘蔗黑穗病抗性分级标准

Table 40 Classification criteria for sugarcane smut resistance

抗性等级 Resistance grade	发病率 Incidence (%)		抗性类型 Resistance types
	新植蔗 New plant	宿根蔗 Ratoon	
1	0～3	0～6	高抗 (high resistance, HR)
2	4～6	7～12	抗 (resistance, R)
3	7～9	13～16	抗 (resistance, R)
4	10～12	17～20	中抗 (moderate resistance, MR)
5	13～25	21～30	中感 (moderate susceptibility, MS)
6	26～35	31～40	感 (susceptibility, S)
7	36～50	41～60	感 (susceptibility, S)
8	51～75	61～80	高感 (high susceptibility, HS)
9	76～100	81～100	高感 (high susceptibility, HS)

5.5 甘蔗生物固氮机理及氮高效利用 Mechanism of Biological Nitrogen Fixation and Efficient Nitrogen Utilization in Sugarcane

5.5.1 甘蔗生物固氮机理 Mechanism of Biological Nitrogen Fixation in Sugarcane

Impact of sugarcane–legume intercropping on diazotrophic microbiome

The present study discussed the application of the intercropping system to improve land use efficacy and soil microbial activity. We assessed linkages of soil properties and unculturable diazotrophs community under three cultivation systems (monoculture sugarcane, peanut–sugarcane and soybean–

sugarcane intercropping). Rhizosphere soil of sugarcane was sampled and DNA was extracted. We amplified the nifH gene and sequenced by high throughput sequencing. The bioinformatics analysis of sequenced data obtained a total of 436 458 nifH gene reads that are classified into 3 201 unique operational taxonomic units (OTUs). A higher percentage of exclusive OTUs identified under soybean–sugarcane intercropping (<375). The microbial structure results showed that Alphaproteobacteria and Beta-proteobacteria were the dominant groups in all three cultivation systems. While genus such as Bradyrhizobium, Burkholderia, Pelomonas, and Sphingomonas was predominant in the intercropping systems and these diazotrophic bacterial communities were positively correlated to the soil pH and soil enzyme protease. Additionally, a lower quantity of available P in the soil of intercrops indicated a strong link between soil nutrients uptake and microbial activity. The results of the present study concluded some interesting facts of intercropping systems that positively improved the soil microbial activity and this kind of strategy could help to cultivate multiple crops to improve the economic growth of the country by sustainable sugarcane production.

(Manoj Kumar Solanki, Fei-yong Wang, Chang-ning Li, Zhen Wang, Tao-ju Lan, Rajesh Kumar Singh, Pratiksha Singh, Li-tao Yang, Yang-rui Li*)

5.5.2 甘蔗氮高效利用 Efficient Nitrogen Utilization in Sugarcane

施氮水平对不同甘蔗品种产量和蔗糖分的影响

采用裂区设计，以6个甘蔗栽培品种（桂选B9、桂斐1号、新台糖22号、桂糖42号、桂糖46号和桂糖47号）为主区，以尿素施用量（150 kg/hm² 和600 kg/hm²）为副区，于伸长期施入全部氮肥，测定施氮水平对甘蔗农艺性状及产量和糖分的影响（表41）。研究发现不同氮素水平对甘蔗萌芽出苗影响不大，而分蘖率随施氮量增加而提高；同一品种的株高和茎径在低氮和高氮处理间差异明显，而除了桂糖42号，其他品种高氮处理有效茎数比低氮处理多；不同品种间的株高、茎径和有效茎数存在显的差异；试验所有品种甘蔗产量随着尿素施用量在150 kg/hm² 到600 kg/hm² 升高而增加；桂糖42号的高氮处理蔗糖分高于低氮，其他品种甘蔗蔗糖分均是低氮处理高于高氮处理；氮肥施用增加对增加甘蔗分蘖率、有效茎数和甘蔗产量有明显的效应，但降低甘蔗蔗糖分；不同品种对氮肥的响应不同，与品种的特性有很大关系。因此，生产上应结合甘蔗品种特性，有针对性施用氮肥，促进甘蔗的高效节本栽培。

（范业赓，丘立杭，陈荣发，周慧文，黄杏，卢星高，甘崇琨，吴建明*，李杨瑞*）

Effects of nitrogen application level on yield and sucrose content of different sugarcane cultivars

The experiment was designed with a split zone, and six sugarcane cultivars (GXB9, GF1, ROC22, GT42, GT46 and GT47) were used as the main plot, and the application amount of urea (150 kg/hm² and 600 kg/hm²) was the secondary plot, and all the nitrogen fertilizers were applied on June 10, 2016. The effects of nitrogen application on the agronomic traits and sugar content were determined (Table 41). The results indicated that nitrogen had little effect on seedling emergence, but it enhanced the tillering; within each cultivar, no significant difference for plant height and stalk diameter between two treatments was observed. Except for GT42, the millable stalks was higher in the high nitrogen treatment than the low nitrogen treatment; cultivar differences were significant for plant height, stalk diameter and millable stalks. The cane yield in all sugarcane cultivars increased with the increased urea rates from 150 kg/hm^2 to 600 kg/hm^2, however, an opposite effect was observed for sucrose content except for the cultivar GT42. In conclusion, increasing nitrogen fertilizer rates increased the tillering rate, millable stalks and cane yield, but decreased the sucrose content in sugarcane. Different response of cultivar to nitrogen suggested to take into account the cultivar characteristics in sugarcane production.

(Ye-geng Fan, Li-hang Qiu, Rong-fa Chen, Hui-wen Zhou, Xing Huang, Xing-gao Lu, Chong-kun Gan, Jian-ming Wu[*], Yang-rui Li[*])

表41 施氮水平对不同甘蔗品种农艺性状的影响

Table 41 Effects of nitrogen treatments on agronomic traits of different sugarcane varieties

处理 Treatment	出苗率（%） Emergence rate	分蘖率（%） Tillering rate	株高（cm） Plant height	茎径（cm） Stalk diameter	有效茎数（条/hm^2） Millable stalk
A1B1	78.60 abAB	11.33 dB	315.73 bcBCD	2.52 cDE	62 412 aA
A1B2	75.79 abcABC	14.68 dB	317.08 bABC	2.45 cdEF	63 127 aA
A2B1	65.09 eC	11.51 dB	294.39 efEF	2.66 bABC	37 082 eDE
A2B2	71.93 abcdeABC	14.91 dB	300.16 defCDEF	2.63 bBCD	42 879 cdeBCDE
A3B1	70.96 bcdeABC	29.76 abcAB	299.13 defDEF	2.28 eG	41 688 cdeCDE
A3B2	64.91 eC	29.80 abcAB	289.65 fF	2.38 dFG	46 531 bcdBCD
A4B1	74.47 abcdABC	34.49 abA	317.71 bAB	2.65 bBCD	52 487 bB
A4B2	79.65 aA	38.04 aA	333.50 aA	2.53 cCDE	51 693 bB
A5B1	72.63 abcdeABC	20.39 cdAB	309.63 bcdBCDE	2.78 aA	46 452 bcdBCDE
A5B2	71.22 bcdeABC	22.62 bcdAB	310.08 bcdBCDE	2.68 bAB	48 040 bcBC
A6B1	67.63 deBC	11.57 dB	304.54 cedBCDEF	2.64 bBCD	36 844 eE
A6B2	68.77 cdeABC	15.08 dB	315.23 bcBCD	2.70 abAB	40 099 deCDE

注：同列数据后不同大、小写字母分别表示差异达极显著（$P<0.01$）和显著水平（$P<0.05$），下同。

Note: uppercase and lowercase indicate difference is significant at the 0.01 and 0.05 levels, respectively, hereinafter.

不同施氮水平下固氮菌肥对甘蔗的应用效果试验

探讨固氮菌肥在不同施氮水平下对甘蔗产量和蔗糖品质的影响，明确固氮菌肥的最佳施氮范围，试验设5个处理，CK为常规尿素追量900 kg/hm^2，固氮菌浸种的A、B、C、D处理中，尿素追量分别为630 kg/hm^2、540 kg/hm^2、450 kg/hm^2、360 kg/hm^2；调查各处理的甘蔗农艺性状、产量性状、蔗茎产量及蔗糖品质表现（表42）。与CK相比，A、B、C、D处理下，甘蔗分蘖率减少9.72%～16.61%、并随施氮量的减少而减少、差异显著。同时甘蔗产量增加0.58%～5.34%、处理间差异不明显；尿素追量540～630 kg/hm^{2}内，固氮菌肥对甘蔗施用的综合效果最佳。固氮菌肥与适量氮肥配施，可以提高肥料利用率、延长肥料作用时间，促进甘蔗生长，增加和稳定产量，减少化肥施量，降低生产成本和增加收入，达到减肥增效的效果。

(梁阗，何为中，谭宏伟，高轶静，庞天，李德伟，覃振强)

(续)

种植模式 Cropping pattern(C)	施氮水平 (kg/hm^2) Nitrogen level(N)	出苗数 (×10^4/hm^2) Emergency number	分蘖数 (×10^4/hm^2) Tiller number	有效茎数 (×10^4/hm^2) Number of millable stalks	成茎率（%） Percentage of millable stalks	氮素吸收量 (kg/hm^2) Nitrogen uptake	蔗茎产量 (t/hm^2) Cane yield
甘蔗-绿豆 Sugarcane-mungbean	0	8.68±0.11b	7.89±0.15d	5.43±0.04b	32.80±0.67ab	84.29±5.74d	70.12±1.84d
	231	8.70±0.18b	10.79±0.10b	6.72±0.12a	34.49±0.44a	143.18±2.36b	99.00±1.94b
	330	8.75±0.22b	10.88±0.09b	6.84±0.26a	34.81±1.16a	188.61±5.07a	122.44±3.64a
F	C	55.51**	139.31**	23.41*	123.65**	125.89**	61.62**
	N	0.04	425.98**	86.50**	15.30**	271.26**	174.21**
	C×N	0.06	1.12	6.85*	5.98*	6.14*	0.87

注：同列数据后的不同小写字母差异显著（$P<0.05$, Duncan's 法）；"*" 和 "**" 分别表示影响达到0.05和0.01的显著水平。

Note: Different lowercase letters in the same column indicate significant difference ($P<0.05$, Duncan's method); "*"and"**"indicate that the effect reaches 0.05 and 0.01 significance levels respectively.

苗期甘蔗氮高效基因型评价指标的筛选

本研究通过低氮压力选择，筛选出甘蔗氮高效种质，分析影响甘蔗氮高效的重要指标，为甘蔗氮高效育种及栽培提供理论依据（图23）。以58份甘蔗种质资源为材料，在苗期采用正常供氮（2 mmol/L N）和低氮（0.2 mmol/L N）处理，分析甘蔗植株形态、干重及氮素在各器官中累积分配的特征。通过主成分分析方法筛选影响甘蔗氮高效利用的重要指标，通过聚类分析对58份种质进行聚类。结果表明，低氮（0.2 mmol/L N）处理可以明显从植物形态区分不同种质的氮利用差异，58份种质低氮条件下的干重范围在0.64～14.75 g/株，氮累积量在5.53～63.00 mg/株，氮利用率范围在115.40～279.30 g/kg之间。对低氮压力下甘蔗干重及氮累积等25个指标进行成分主成分分析后，提取出4个主要成分，总贡献率为92.35%。通过高、低氮条件下与氮利用效率有关的氮转移系数及基因潜力等19个指标分析后提取出5个成分主成分，总贡献率为82.21%。影响甘蔗氮高效的重要指标有甘蔗的干重（全株，叶，根）、氮累积量（全株，叶，茎）、氮利用率（全株，叶）、叶的相对氮利用率、茎的基因潜力、茎的相对干物质量和茎的相对氮累积量。经聚类分析后初步将58份甘蔗种质分为氮高效基因型、偏氮高效基因型、偏氮低效基因型和氮低效基因型。

（杨柳，廖芬*，Muhammad Anas，李强，彭李顺，黄东亮，李杨瑞*）

Screening of sugarcane with high nitrogen efficiency at seedling stage

The purpose of this experiment is to screen high nitrogen use efficiency (NUE) genotypes under low nitrogen pressure selection system (Fig.23). Furthermore, the important indicesindices affecting the NUE of sugarcane were analyzed. The results could provide a theoretical basis for the high NUE breeding and cultivation of sugarcane. In this study, seedlings of 58 sugarcane genotypes were evaluated in a hydroponic experiment with low-N (0.2 mM N) and normal-N (2 mmol/L N) treatment. The growth, dry biomass and N cumulative and distribution characteristic in various organs were evaluated according to descriptive statistic, principal component analysis and cluster analysis .The results indicated that morphology, biomass, Nitrogen efficiency traits show high genotypic variation for N treatments. Under low-N treatment, the dry weight of 58 genotypes varied from 0.64 ~ 14.75 g/plant, nitrogen accumulation was from 5.53 ~ 63.00 mg/plant and NUE was from 115.40 ~ 279.30 g/kg. Four factors were extracted with variance contribution approximated to 92.35%, according to 25 parameters such as dry weight and N uptake et al with N deficiency. And another five factors were extracted with variance contribution approximated to 82.21%, according to 19 parameters under both normal and low N treatment such as Nitrogen transfer coefficient, genetic potential et al. The data revealed that dry weight(whole plant, leaf, root), N uptake(whole plant, leaf, shoot), NUE (whole plant, leaf), leaf relative NUE, shoot relative dry weight, shoot relative N uptake, shoot genetic potential were the key factors that involved in sugarcane high NUE. 58 sugarcane genotypes were cluster into four group: high NUE group, slightly higher NUE group, slightly lower NUE group and low group.

(Liu Yang, Fen Liao[*], Muhammad Anas, Qiang Li, Li-shun Peng, Dong-liang Huang, Yang-rui Li[*])

图 23 不同氮水平下甘蔗苗期生长情况对比（a 为正常供氮处理，b 为低氮处理）

Fig. 23 Comparison for seedling growth of sugarcane under different nitrogen levels (a is normal N treatment, b is N deficiency treatment)

施用生物炭对甘蔗土壤化学性质及氮损失的影响

文章研究的目的是分析生物炭对甘蔗不同生长期土壤大量元素 N、P、K 及有机质含量、pH、电导率和土壤中氮损失的影响，为甘蔗生产上应用生物炭提供理论基础（图24）。该试验采用桶栽甘蔗，施用不同用量的木薯茎秆生物炭，用 ^{15}N 示踪方法分析 N 含量，并测定 P、K 及有机质含量、pH、电导率等指标。试验结果发现，生物炭处理可以明显增加甘蔗分蘖期土壤中的 N 含量，与不施生物炭对照相比增加了 4.17% ~ 33.33%，同时 N 损失从 48.85% 降至 31.26%；P 含量在分蘖期和成熟期也分别比对照增加了 13.19% ~ 61.67%，并促进了中后期土壤有机质含量增加，促进成熟期K在土壤中保留，显著提高了土壤的pH，但对电导率无显著的影响。生物炭处理增加土壤N、P、K元素的作用与提高土壤的pH值和有机质含量有关。生物炭施用可改善甘蔗土壤的营养元素含量，增加N、P、K在土壤中的保留，提高土壤肥力，可以作为一种土壤改良剂在甘蔗生产中应用。

（廖芬，桂杰，杨柳*，李强，Muhammad Anas，李杨瑞*）

Effects of biochar application on soil chemical properties and nitrogen loss of sugarcane

The purpose of this study was to analyze the effects of biochar on soil N, P, K and organic matter content, pH value, electrical conductivity and nitrogen loss in sugarcane at different growth stages, so as to provide a theoretical basis for the application of biochar in sugarcane production (Fig. 24). In this experiment, sugarcane was planted in buckets and different amounts of cassava stem biochar were applied. ^{15}N tracer method was used to analyze the content of N, P, K, organic matter content, pH value, electrical conductivity and so on. The results showed that biochar treatment could significantly increase the content of N in sugarcane soil at tillering stage, which increased by 4.17% ~ 33.33% compared with the control without biochar, and the N loss decreased from 48.85% to 31.26%. P content in tillering stage and mature stage also increased by 13.19% ~ 61.67% compared with the control, and promoted the increase of soil organic matter content in the middle and later stage, promoted K retention in soil at mature stage, and significantly increased soil pH value, but had no significant effect on electrical conductivity. The effect of biochar treatment on increasing soil N, P and K elements is related to the increase of soil pH value and organic matter content. The application of biochar can improve the content of nutrient elements in sugarcane soil, increase the retention of N, P and K in soil, and improve soil fertility, which can be used as a soil improver in sugarcane production.

(Fen Liao, Jie Gui, Liu Yang*, Qiang Li, Muhammad Anas, Yang-rui Li*)

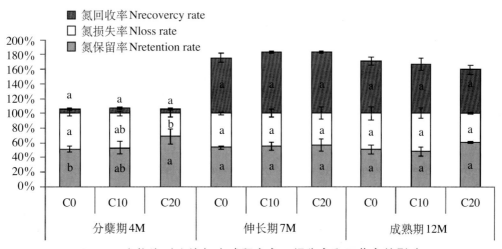

图24　生物炭对土壤氮素残留离率、损失率和回收率的影响

Fig. 24　Effect of biochar on N retention rate, N loss rate and N recovery rate in soil

不同生物质来源生物炭品质的因子分析与综合评价

建立一套适合生物炭品质评价的方法，探求影响生物炭品质的主要影响因子（表24）。选

用8种不同生物质材料,在3种温度下制备并获得24种生物炭材料(Y1~Y24),测定16项相关品质指标,采用隶属函数法对各项指标数据进行转化,采用软件SPSS19.0进行因子分析,采用四次方最大旋转法获得因子载荷矩阵,计算样品每个公因子分值与相应权重之积的累加和,得到综合评价分值。对16项指标进行相关性分析和因子分析,建立基于因子分析的生物炭品质的综合评价体系,并根据综合评价得分对生物炭进行优良度排序。24种生物炭16个品质指标经因子分析,提取了5个特征根>1的公因子,累计方差贡献率达到77.910%,第1公因子以C含量、阳离子交换量(CEC)和pH贡献率较大,达到31.090%,第2公因子以比表面积和孔容贡献率较大,达到19.878%,第3公因子以H原子含量贡献率较大,达到12.819%,第4公因子以磷、钾含量贡献率较大,达到7.479%,第5公因子以NH_4^+-N吸附量贡献率较大,达到6.643%。因子分析方法可以作为评价生物炭品质的方法,根据因子分析评价方法,确定影响生物炭品质最关键的因子是化学性质因子(C含量、C/N比、C/H比、pH、CEC)、物理性质因子(比表面积和孔容)、活化能量因子(H原子含量)、营养因子(P和K含量)和氨态氮吸附能力因子。

(廖芬,杨柳[*],李强,Muhammad Anas,薛进军,黄东亮,李杨瑞[*])

Factor analysis and comprehensive evaluation for quality of biochar derived from different biomass

This work was conducted to establish a method suitable for quality evaluation of biochar and to explore the main factors influencing biochar quality (Table 44). Toally 24 biochars (Y1–Y24) were prepared from eight biomass materials under three different temperatures. Sixteen quality indices of biochars were measured, and the data were converted using the subordinate function. SPSS 19.0 was used for factor analysis. Factor loading matrix was obtained using the biquadratic maximum rotation method. The comprehensive evaluation scores were calculated. Correlation analysis and factor analysis were performed for the 16 indicesindices, and the comprehensive evaluation system of biochar quality was established based on factor analysis. The biochars were ranked using the comprehensive evaluation scores. Through factor analysis of 16 quality indices of 24 biochars, we extracted five common factors with eigenvalues above one and their cumulative variance contribution was 77.910%. The first common factor consisted of carbon content, cation exchange capacity (CEC) and pH, with variance contribution of 31.090%. The second common factor consisted of specific surface area and pore volume, with variance contribution of 19.878%. The third common factor consisted of hydrogen atom content with variance contribution of 12.819%. The fourth common factor consisted of phosphorus and potassium contents with variance contribution of 7.479%. The fifth common factor consisted of maximum adsorption capacity of NH^{4+}-N with variance contribution of 6.643%. Factor analysis is a good statistical method for evaluating biochar quality. The key factors affecting biochar quality include chemical characteristic factors (C content, C/N ratio, C/H ratio, pH and CEC), physical characteristic factors (specific surface area and pore volumc), active energy factor (H atom content), nutrition factor (P and K contents) and maximum adsorption capacity of NH_4^+-N.

(Fen Liao, Liu Yang[*], Qiang Li, Muhammad Anas, Jin-jun Xue, Dong-liang Huang, Yang-rui Li[*])

表44 种生物炭的综合得分及排名

Table 44　Comprehensive scores and ranking of 24 biochars

排名 Ranking	因子1		因子2		因子3		因子4		因子5		综合得分 Comprehensive score (Dn)	生物炭材料编号 Biochar number
	得分 Score (F)	生物炭材料编号 Biochar number	得分 Score (F)	生物炭材料编号 Biochar number	得分 Score (F)	生物炭材料编号 Biochar number	得分 Score (F)	生物炭材料编号 Biochar number	得分 Score (F)	生物炭材料编号 Biochar number		
1	1.30	Y9	2.13	Y24	1.72	Y10	2.28	Y2	1.44	Y6	0.69	Y24
2	1.24	Y15	2.04	Y22	1.60	Y13	1.31	Y20	1.44	Y4	0.37	Y6
3	1.03	Y12	1.72	Y23	1.16	Y12	1.02	Y17	1.29	Y5	0.36	Y23
4	1.02	Y6	1.31	Y3	0.85	Y23	0.92	Y13	1.24	Y24	0.36	Y17
5	1.00	Y18	1.27	Y21	0.71	Y17	0.89	Y22	0.83	Y22	0.34	Y18
6	0.87	Y11	0.79	Y20	0.61	Y22	0.75	Y19	0.82	Y7	0.33	Y15
7	0.75	Y17	0.17	Y2	0.42	Y16	0.66	Y11	0.80	Y3	0.32	Y3
8	0.70	Y24	0.10	Y18	0.42	Y5	0.52	Y3	0.58	Y2	0.30	Y20
9	0.55	Y21	0.02	Y19	0.41	Y11	0.45	Y15	0.57	Y11	0.27	Y11
10	0.49	Y5	−0.03	Y1	0.33	Y18	0.35	Y9	0.36	Y10	0.21	Y21
11	0.47	Y3	−0.06	Y6	0.26	Y6	0.14	Y5	0.34	Y17	0.21	Y12
12	0.43	Y8	−0.23	Y15	0.23	Y24	0.12	Y10	0.06	Y14	0.12	Y5
13	0.37	Y20	−0.33	Y17	0.13	Y15	0.08	Y14	−0.06	Y18	0.05	Y9
14	0.24	Y14	−0.44	Y14	0.03	Y21	−0.05	Y8	−0.07	Y8	0.04	Y22
15	0.08	Y23	−0.52	Y9	−0.07	Y14	−0.34	Y12	−0.35	Y13	−0.01	Y14
16	−0.37	Y13	−0.70	Y10	−0.13	Y19	−0.35	Y18	−0.56	Y23	−0.09	Y13
17	−0.68	Y4	−0.71	Y11	−0.14	Y20	−0.57	Y16	−0.59	Y16	−0.25	Y10
18	−1.04	Y2	−0.73	Y8	−0.21	Y4	−0.87	Y6	−0.82	Y20	−0.25	Y2
19	−1.15	Y10	−0.77	Y16	−0.44	Y1	−0.87	Y24	−0.83	Y12	−0.27	Y8
20	−1.18	Y16	−0.90	Y12	−1.30	Y2	−0.96	Y1	−0.93	Y15	−0.44	Y4
21	−1.34	Y7	−0.92	Y5	−1.38	Y3	−1.03	Y23	−1.17	Y9	−0.49	Y19
22	−1.39	Y19	−0.97	Y4	−1.58	Y9	−1.33	Y4	−1.36	Y1	−0.55	Y16
23	−1.54	Y1	−1.11	Y13	−1.68	Y7	−1.53	Y21	−1.47	Y19	−0.70	Y1
24	−1.83	Y22	−1.12	Y7	−1.97	Y8	−1.59	Y7	−1.55	Y21	−0.92	Y7

6 附 录
APPENDIX

6.1 实验室学术委员会和固定人员组成
Academic Committee and Staff

表45 广西甘蔗遗传改良重点实验室学术委员会
Table 45 Academic Committee of Guangxi Key Laboratory of Sugarcane Genetic Improvement

姓 名 Name	职称 Title	专业 Speciality	工作单位 Work unit	学委会职务 Duty in committee
彭 明	研究员	分子遗传学与基因组学	中国热带农科院热带生物技术所	主任委员
李杨瑞	教 授	甘蔗育种及栽培	广西农业科学院	副主任委员
谭宏伟	研究员	土壤农化	广西农科院甘蔗研究所	委员
许莉萍	研究员	作物学	福建农林大学	委员
张树珍	研究员	作物遗传育种	中国热带农科院热带生物技术所	委员
沈万宽	研究员	甘蔗抗病育种	华南农业大学	委员
郭家文	研究员	甘蔗栽培	云南农业科学院甘蔗研究所	委员
齐永文	研究员	植物分子育种	广东省生物工程研究所	委员
董登峰	教 授	植物学	广西大学	委员
黄东亮	研究员	甘蔗生物技术	广西农业科学院甘蔗研究所	委员
周 会	研究员	甘蔗遗传育种	广西农业科学院甘蔗研究所	委员

表46 广西甘蔗遗传改良重点实验室主要人员
Table 46 Staff in Guangxi Key Laboratory of Sugarcane Genetic Improvement

序号 Number	姓 名 Name	性别 Gender	出生年月 Birth	学位 Degree	职称 Title	专业/研究方向 Speciality	博导 Ph.D. Tutor	硕导 M.S. Tutor	备注 Comment
1	李杨瑞	男	1957-04	博士	教授	甘蔗栽培与遗传育种			学术带头人
2	谭宏伟	男	1961-02	学士	研究员	土壤农业化学			学术带头人
3	杨丽涛	女	1961-02	博士	教 授	甘蔗功能基因组			研究骨干
4	黄东亮	男	1973-11	博士	研究员	甘蔗功能基因组			学术带头人
5	杨荣仲	男	1963-09	博士	研究员	甘蔗遗传育种			学术带头人
6	吴建明	男	1978-04	博士	研究员	甘蔗功能基因组			研究骨干

(续)

序号 Number	姓名 Name	性别 Gender	出生年月 Birth	学位 Degree	职称 Title	专业/研究方向 Speciality	博导 Ph.D. Tutor	硕导 M.S. Tutor	备注 Comment
7	王维赞	男	1970-04	硕士	研究员	甘蔗栽培生理			学术带头人
8	王伦旺	男	1965-09	学士	研究员	甘蔗遗传育种			学术带头人
9	张革民	男	1968-09	学士	研究员	甘蔗育种及种质资源创新			学术带头人
10	刘昔辉	男	1982-09	硕士	研究员	甘蔗育种及种质资源创新			管理人员
11	吴凯朝	男	1979-06	博士	副研究员	甘蔗分子育种			管理人员
12	林 丽	女	1982-04	博士	副研究员	甘蔗固氮			管理人员
13	廖 芬	女	1978-08	硕士	副研究员	甘蔗分子育种			管理人员
14	黄 杏	女	1984-01	博士	副研究员	作物栽培生理及分子育种			管理人员
15	汪 淼	女	1982-10	硕士	副研究员	甘蔗分子育种			管理人员
16	桂意云	女	1980-02	硕士	助理研究员	甘蔗分子育种			管理人员
17	陈忠良	男	1984-03	硕士	助理研究员	生物化学与分子生物			管理人员
18	莫璋红	女	1976-08	硕士	助理研究员	甘蔗栽培生理			管理人员
19	秦翠鲜	女	1984-11	硕士	助理研究员	甘蔗分子育种			管理人员
20	丘立杭	男	1983-05	硕士	副研究员	甘蔗栽培生理及分子生物学			管理人员
21	陈荣发	女	1986-08	硕士	助理研究员	甘蔗栽培生理及分子生物学			管理人员
22	徐 林	女	1983-11	硕士	助理研究员	甘蔗育种			管理人员
23	范业赓	男	1979-10	硕士	助理研究员	甘蔗栽培生理			管理人员
24	邓智年	女	1976-06	博士	副研究员	甘蔗分子育种			研究骨干
25	周 会	男	1976-05	博士	研究员	甘蔗育种及种质资源创新			研究骨干
26	李 松	男	1964-03	硕士	研究员	甘蔗健康种苗培育			研究骨干
27	游建华	男	1964-07	硕士	研究员	甘蔗育种			研究骨干
28	黄诚华	男	1974-02	博士	副研究员	甘蔗植保			研究骨干
29	何为中	男	1965-06	硕士	研究员	甘蔗生物技术			研究骨干
30	李长宁	男	1984-10	博士	副研究员	甘蔗栽培生理及分子生物学			研究骨干
31	宋修鹏	男	1984-09	博士	助理研究员	甘蔗植保			研究骨干
32	罗 霆	女	1980-11	博士	副研究员	甘蔗育种及种质资源创新			研究骨干
33	高轶静	女	1981-06	硕士	副研究员	甘蔗育种及种质资源创新			研究骨干
34	张保青	男	1980-09	博士	助理研究员	甘蔗种质资源			研究骨干
35	唐仕云	男	1978-03	硕士	副研究员	甘蔗遗传育种			研究骨干
36	李 鸣	男	1977-09	博士	研究员	甘蔗遗传育种			研究骨干
37	谭 芳	女	1964-03	学士	副研究员	甘蔗遗传育种			研究骨干
38	梁 强	男	1981-04	硕士	副研究员	甘蔗栽培生理			研究骨干
39	谢金兰	女	1977-01	硕士	副研究员	甘蔗栽培生理			研究骨干
40	覃振强	男	1975-06	博士	副研究员	甘蔗植保			研究骨干

(续)

序号 Number	姓名 Name	性别 Gender	出生年月 Birth	学位 Degree	职称 Title	专业/研究方向 Speciality	博导 Ph.D. Tutor	硕导 M.S. Tutor	备注 Comment
41	林善海	男	1979-08	博士	副研究员	甘蔗植保			研究骨干
42	刘俊仙	女	1982-10	硕士	副研究员	甘蔗生物技术			研究骨干
43	段维兴	男	1984-08	硕士	副研究员	甘蔗育种及种质资源创新			研究骨干
44	王泽平	男	1983-06	博士	副研究员	甘蔗育种及种质资源创新			研究骨干
45	周忠凤	女	1974-06	学士	高级农艺师	甘蔗遗传育种			研究骨干
46	贤 武	女	1971-11	学士	副研究员	甘蔗遗传育种			研究骨干
47	廖江雄	男	1964-09	博士	副研究员	甘蔗遗传育种			研究骨干
48	李 翔	男	1981-10	博士	副研究员	甘蔗遗传育种			研究骨干
49	雷敬超	男	1981-08	硕士	助理研究员	甘蔗遗传育种			研究骨干
50	黄海荣	男	1982-10	硕士	助理研究员	甘蔗遗传育种			研究骨干
51	张荣华	男	1982-09	硕士	助理研究员	甘蔗生物技术			研究骨干
52	庞 天	男	1968-06	大专	副研究员	甘蔗新品种推广			研究骨干
53	樊保宁	男	1963-10	硕士	副研究员	甘蔗栽培生理			研究骨干
54	罗亚伟	男	1968-01	学士	助理研究员	甘蔗栽培生理			研究骨干
55	梁 阗	男	1963-10	大专	农艺师	甘蔗栽培生理			研究骨干
56	刘晓燕	女	1982-04	硕士	助理研究员	甘蔗栽培生理			研究骨干
57	李毅杰	男	1984-11	硕士	助理研究员	甘蔗栽培生理			技术人员
58	潘雪红	女	1979-05	硕士	助理研究员	甘蔗植保			技术人员
59	韦金菊	女	1981-09	硕士	助理研究员	甘蔗植保			技术人员
60	商显坤	男	1983-07	硕士	助理研究员	甘蔗植保			技术人员
61	魏吉利	女	1979-11	硕士	助理研究员	甘蔗植保			技术人员
62	翁梦苓	女	1982-10	硕士	助理研究员	甘蔗栽培生理			技术人员
63	刘红坚	女	1976-12	学士	助理研究员	甘蔗生物技术			技术人员
64	刘丽敏	女	1978-02	硕士	助理研究员	甘蔗生物技术			技术人员
65	卢曼曼	女	1983-09	硕士	助理研究员	甘蔗生物技术			技术人员
66	杨翠芳	女	1985-03	硕士	助理研究员	甘蔗育种及种质资源创新			技术人员
67	经 艳	女	1982-04	硕士	助理研究员	甘蔗遗传育种			技术人员
68	邓宇驰	男	1984-10	硕士	农艺师	甘蔗遗传育种			技术人员
69	周 珊	女	1985-07	硕士	助理研究员	甘蔗育种及种质资源创新			技术人员
70	李德伟	男	1980-08	硕士	副研究员	甘蔗植保			技术人员
71	张小秋	女	1987-04	硕士	助理研究员	甘蔗植保			技术人员
72	周慧文	女	1990-08	硕士	助理研究员	甘蔗栽培生理			技术人员
73	颜梅新	男	1978-04	博士	副研究员	甘蔗植保			技术人员
74	闫海锋	男	1980-09	博士	助理研究员	甘蔗分子育种			技术人员

(续)

序号 Number	姓名 Name	性别 Gender	出生年月 Birth	学位 Degree	职称 Title	专业/研究方向 Speciality	博导 Ph.D. Tutor	硕导 M.S. Tutor	备注 Comment
75	黄玉新	女	1989-08	硕士	助理研究员	甘蔗种质资源创新利用			技术人员
76	李傲梅	女	1995-03	硕士	助理研究员	甘蔗植保			技术人员

表47　农业农村部广西甘蔗生物技术与遗传改良重点实验室学术委员会
Table 47　Acadamic Committee of Key Laboratory of Sugarcane Biotechnology and Genetic Improvement (Guangxi), Ministry of Agriculture and Rural Affairs, P.R.China

姓名 Name	职称 Title	专业 Speciality	工作单位 Work unit	学委会职务 Duty in committee
彭　明	研究员	分子遗传学与基因组学	中国热带农业科学院热带生物技术研究所	主任委员
李杨瑞	教　授	甘蔗育种及栽培	广西农业科学院	副主任委员
谭宏伟	研究员	土壤农化	广西农业科学院甘蔗研究所	委员
许莉萍	研究员	作物学	福建农林大学	委员
张树珍	研究员	作物遗传育种	中国热带农业科学院热带生物技术研究所	委员
沈万宽	研究员	甘蔗抗病育种	华南农业大学	委员
郭家文	研究员	甘蔗栽培	云南省农业科学院甘蔗研究所	委员
齐永文	研究员	植物分子育种	广东省生物工程研究所	委员
董登峰	教　授	植物学	广西大学	委员
黄东亮	研究员	甘蔗生物技术	广西农业科学院甘蔗研究所	委员
周　会	研究员	甘蔗遗传育种	广西农业科学院甘蔗研究所	委员

表48　农业部广西甘蔗生物技术与遗传改良重点实验室主要人员
Table 48　Staff of Key Laboratory of Sugarcane Biotechnology and Genetic Improvement (Guangxi), Ministry of Agriculture, P.R. China

序号 Number	姓名 Name	性别 Gender	出生年月 Birth	学位 Degree	职称 Title	专业/研究方向 Speciality	博导 Ph.D. Tutor	硕导 M.S. Tutor	备注 Comment
1	李杨瑞	男	1957-04	博士	教授	甘蔗栽培与遗传育种			学术带头人
2	谭宏伟	男	1961-02	学士	研究员	土壤农业化学			学术带头人
3	杨丽涛	女	1961-02	博士	教　授	甘蔗功能基因组			研究骨干
4	黄东亮	男	1973-11	博士	研究员	甘蔗功能基因组			学术带头人
5	杨荣仲	男	1963-09	博士	研究员	甘蔗遗传育种			学术带头人
6	吴建明	男	1978-04	博士	研究员	甘蔗功能基因组			研究骨干
7	王维赞	男	1970-04	硕士	研究员	甘蔗栽培生理			学术带头人
8	王伦旺	男	1965-09	学士	研究员	甘蔗遗传育种			学术带头人

(续)

序号 Number	姓名 Name	性别 Gender	出生年月 Birth	学位 Degree	职称 Title	专业/研究方向 Speciality	博导 Ph.D. Tutor	硕导 M.S. Tutor	备注 Comment
9	张革民	男	1968-09	学士	研究员	甘蔗育种及种质资源创新			学术带头人
10	刘昔辉	男	1982-09	硕士	研究员	甘蔗育种及种质资源创新			管理人员
11	吴凯朝	男	1979-06	博士	副研究员	甘蔗分子育种			管理人员
12	林丽	女	1982-04	博士	副研究员	甘蔗固氮			管理人员
13	廖芬	女	1978-08	硕士	副研究员	甘蔗分子育种			管理人员
14	黄杏	女	1984-01	博士	副研究员	作物栽培生理及分子育种			管理人员
15	汪淼	女	1982-10	硕士	副研究员	甘蔗分子育种			管理人员
16	桂意云	女	1980-02	硕士	助理研究员	甘蔗分子育种			管理人员
17	陈忠良	男	1984-03	硕士	助理研究员	生物化学与分子生物			管理人员
18	莫璋红	女	1976-08	硕士	助理研究员	甘蔗栽培生理			管理人员
19	秦翠鲜	女	1984-11	硕士	助理研究员	甘蔗分子育种			管理人员
20	丘立杭	男	1983-05	硕士	副研究员	甘蔗栽培生理及分子生物学			管理人员
21	陈荣发	女	1986-08	硕士	助理研究员	甘蔗栽培生理及分子生物学			管理人员
22	徐林	女	1983-11	硕士	助理研究员	甘蔗育种			管理人员
23	范业赓	男	1979-10	硕士	助理研究员	甘蔗栽培生理			管理人员
24	邓智年	女	1976-06	博士	副研究员	甘蔗分子育种			研究骨干
25	周会	男	1976-05	博士	研究员	甘蔗育种及种质资源创新			研究骨干
26	李松	男	1964-03	硕士	研究员	甘蔗健康种苗培育			研究骨干
27	游建华	男	1964-07	硕士	研究员	甘蔗育种			研究骨干
28	黄诚华	男	1974-02	博士	副研究员	甘蔗植保			研究骨干
29	何为中	男	1965-06	硕士	研究员	甘蔗生物技术			研究骨干
30	李长宁	男	1984-10	博士	副研究员	甘蔗栽培生理及分子生物学			研究骨干
31	宋修鹏	男	1984-09	博士	助理研究员	甘蔗植保			研究骨干
32	罗霆	女	1980-11	博士	副研究员	甘蔗育种及种质资源创新			研究骨干
33	高轶静	女	1981-06	硕士	副研究员	甘蔗育种及种质资源创新			研究骨干
34	张保青	男	1980-09	博士	助理研究员	甘蔗种质资源			研究骨干
35	唐仕云	男	1978-03	硕士	副研究员	甘蔗遗传育种			研究骨干
36	李鸣	男	1977-09	博士	研究员	甘蔗遗传育种			研究骨干
37	谭芳	女	1964-03	学士	副研究员	甘蔗遗传育种			研究骨干

(续)

序号 Number	姓名 Name	性别 Gender	出生年月 Birth	学位 Degree	职称 Title	专业/研究方向 Speciality	博导 Ph.D. Tutor	硕导 M.S. Tutor	备注 Comment
38	梁 强	男	1981-04	硕士	副研究员	甘蔗栽培生理			研究骨干
39	谢金兰	女	1977-01	硕士	副研究员	甘蔗栽培生理			研究骨干
40	覃振强	男	1975-06	博士	副研究员	甘蔗植保			研究骨干
41	林善海	男	1979-08	博士	副研究员	甘蔗植保			研究骨干
42	刘俊仙	女	1982-10	硕士	副研究员	甘蔗生物技术			研究骨干
43	段维兴	男	1984-08	硕士	副研究员	甘蔗育种及种质资源创新			研究骨干
44	王泽平	男	1983-06	博士	副研究员	甘蔗育种及种质资源创新			研究骨干
45	周忠凤	女	1974-06	学士	高级农艺师	甘蔗遗传育种			研究骨干
46	贤 武	女	1971-11	学士	副研究员	甘蔗遗传育种			研究骨干
47	廖江雄	男	1964-09	博士	副研究员	甘蔗遗传育种			研究骨干
48	李 翔	男	1981-10	博士	副研究员	甘蔗遗传育种			研究骨干
49	雷敬超	男	1981-08	硕士	助理研究员	甘蔗遗传育种			研究骨干
50	黄海荣	男	1982-10	硕士	助理研究员	甘蔗遗传育种			研究骨干
51	张荣华	男	1982-09	硕士	助理研究员	甘蔗生物技术			研究骨干
52	庞 天	男	1968-06	大专	副研究员	甘蔗新品种推广			研究骨干
53	樊保宁	男	1963-10	硕士	副研究员	甘蔗栽培生理			研究骨干
54	罗亚伟	男	1968-01	学士	助理研究员	甘蔗栽培生理			研究骨干
55	梁 阗	男	1963-10	大专	农艺师	甘蔗栽培生理			研究骨干
56	刘晓燕	女	1982-04	硕士	助理研究员	甘蔗栽培生理			研究骨干
57	李毅杰	男	1984-11	硕士	助理研究员	甘蔗栽培生理			技术人员
58	潘雪红	女	1979-05	硕士	助理研究员	甘蔗植保			技术人员
59	韦金菊	女	1981-09	硕士	助理研究员	甘蔗植保			技术人员
60	商显坤	男	1983-07	硕士	助理研究员	甘蔗植保			技术人员
61	魏吉利	女	1979-11	硕士	助理研究员	甘蔗植保			技术人员
62	翁梦苓	女	1982-10	硕士	助理研究员	甘蔗栽培生理			技术人员
63	刘红坚	女	1976-12	学士	助理研究员	甘蔗生物技术			技术人员
64	刘丽敏	女	1978-02	硕士	助理研究员	甘蔗生物技术			技术人员
65	卢曼曼	女	1983-09	硕士	助理研究员	甘蔗生物技术			技术人员
66	杨翠芳	女	1985-03	硕士	助理研究员	甘蔗育种及种质资源创新			技术人员

（续）

序号 Number	姓名 Name	性别 Gender	出生年月 Birth	学位 Degree	职称 Title	专业/研究方向 Speciality	博导 Ph.D. Tutor	硕导 M.S. Tutor	备注 Comment
67	经 艳	女	1982-04	硕士	助理研究员	甘蔗遗传育种			技术人员
68	邓宇驰	男	1984-10	硕士	农艺师	甘蔗遗传育种			技术人员
69	周 珊	女	1985-07	硕士	助理研究员	甘蔗育种及种质资源创新			技术人员
70	李德伟	男	1980-08	硕士	副研究员	甘蔗植保			技术人员
71	张小秋	女	1987-04	硕士	助理研究员	甘蔗植保			技术人员
72	周慧文	女	1990-08	硕士	助理研究员	甘蔗栽培生理			技术人员
73	颜梅新	男	1978-04	博士	副研究员	甘蔗植保			技术人员
74	闫海锋	男	1980-09	博士	助理研究员	甘蔗分子育种			技术人员
75	黄玉新	女	1989-08	硕士	助理研究员	甘蔗种质资源创新利用			技术人员
76	李傲梅	女	1995-03	硕士	助理研究员	甘蔗植保			技术人员

6.2 博士后培养和研究生教育 Postdoctoral Fellow Training and Postgraduate Education

表49　2019年博士后培养一览表

Table 49　Postdoctoral fellow training in 2019

序号 No.	姓名 Name	国籍 Nationality	项目名称 Project title	时间 Time		获博士学位单位和专业 Speciality and institution of Ph.D degree
				起 Start	止 End	
1	Rajesh Kumar Singh	印度	Diversity Of Nitrogen Fixing Microbes And Effect Of Potent Nitrogen Fixing Bacteria In Growth Of Sugarcane Plant	2014-11		印度 Rani Durgavati University
2	Pratiksha Singh	印度	Study on the Interaction Mechanism of Sugarcane and Smut Pathogen	2015-05	2021-01	印度 Punjab Agricultural University
3	Mukesh Kumar Malviya	印度	Effect of N-fertilization on rhizobial actions in sugarcane-associated diazotrophs in sugarcane/soybean intercropping fields	2016-08	2021-03	印度 Timarni College
4	Krishan Kumar Verma	印度	Effect of silicon on growth-productivity in sugarcane under water stress: Physiological and Molecular aspects	2018-05		印度 University of Lucknow
5	Anjney Sharma	印度	Physiological and Molecular basis studies on the interaction between sugarcane and smut	2020-01		印度 Rani Durgavati University Jabalpur

表50 2019年毕业博士研究生一览表
Table 50 Ph D. graduated in 2019

姓名 Name	性别 Gender	专业 Speciality	论文题目 Dissertation	导师 Ph.D. Tutor
李翔	男	作物栽培学与耕作学	甘蔗抗倒伏性评价及抗性机制研究 Evaluation on lodging resistance and mechanism of lodging resistance in sugarcane	杨丽涛

表51 2019年毕业硕士研究生一览表
Table 51 Masters graduated in 2019

姓名 Name	性别 Gender	专业 Speciality	论文题目 Thesis	导师 Tutor
何洪良	男	农业硕士	不同抗旱剂浸种对甘蔗"健康种子"抗旱性的影响 Effects on the drought resistance of sugarcane plene soaked with different drought resistant agents	王爱勤 李杨瑞
黄婵	女	作物遗传育种	甘蔗 SoACLA-1 基因遗传稳定性分析及功能验证 Genetic stability analysis and functional verification of sugarcane SoACLA-1 gene	李杨瑞
唐丽华	女	作物遗传育种	转 SoTUA 基因甘蔗T1代的表达分析及功能验证 Expression analysis and functional verification of SoTUA transgenic sugarcane	李杨瑞
毛莲英	女	植物学	固氮菌XD20对不同甘蔗品种的促生效应 Growth promoting effect of nitrogen fixing bacteria XD20 on different sugarcane varieties	邢永秀
刘芳君	女	植物学	甘蔗体细胞融合的分子基础研究 Molecular basis of somatic cell fusion in sugarcane	李素丽
王凡伟	女	植物学	甘蔗ethylene insensitive3-like 1 (ScEIL1)基因分子克隆和表达分析 Molecular cloning and expression analysis of ethylene insensitive3-like 1 (ScEIL1) gene in sugarcane	王爱勤

表52 2019年在读博士研究生一览表
Table 52 Ph.D. postgraduate students in 2019

姓名 Name	性别 Gender	导师 Tutor	入学时间 Entrance date	姓名 Name	性别 Gender	导师 Tutor	入学时间 Entrance date
李健	男	杨丽涛	2014-09	赵培方	男	杨丽涛	2015-09
Aamir Mahood	男	李杨瑞	2016-09	祝开	男	李杨瑞	2017-09
Qaisar Khan	男	李杨瑞	2017-09	Muhammad Anas	男	李杨瑞	2017-09
王震	男	董登峰、李杨瑞	2017-09	郭道君	男	李杨瑞	2018-09
罗培四	男	李素丽	2019-09				

表53 2019年在读硕士研究生一览表

Table 53　M.S. postgraduate students in 2019

姓名 Name	性别 Gender	导师 Tutor	入学时间 Entrance date	姓名 Name	性别 Gender	导师 Tutor	入学时间 Entrance date
陈教云	女	董登峰	2017-09	李 强	男	董登峰	2017-09
韦江璐	女	李杨瑞	2017-09	余卓新	男	李杨瑞	2017-09
黄金玲	女	李素丽	2017-09	梁晓莹	女	李素丽	2017-09
曾许鹏	男	李杨瑞	2017-09	覃 英	女	邢永秀	2017-09
赖沛衡	女	李素丽	2017-09	谢显秋	男	邢永秀	2018-09
谢光杰	男	李杨瑞	2018-09	陆 珍	女	李杨瑞	2018-09
陈炯宇	男	邢永秀	2018-09	黄毓燕	女	邢永秀	2019-09
陆 珍	女	李杨瑞	2018-09	冼春暖	女	董登峰	2019-09
张瑞楠	男	邢永秀	2019-09	陈晓茹	女	李素丽	2019-09
李 浩	男	李素丽	2019-09				

图书在版编目（CIP）数据

广西农业科学院甘蔗发展报告2019：汉英对照 / 农业农村部广西甘蔗生物技术与遗传改良重点实验室，广西甘蔗遗传改良重点实验室编．—北京：中国农业出版社，2021.5

ISBN 978-7-109-28117-2

Ⅰ.①广… Ⅱ.①农…②广… Ⅲ.①甘蔗－栽培技术－汉、英 Ⅳ.①S566.1

中国版本图书馆CIP数据核字（2021）第064370号

GUANGXI NONGYE KEXUEYUAN GANZHE FAZHAN BAOGAO 2019

中国农业出版社出版
地址：北京市朝阳区麦子店街18号楼
邮编：100125
责任编辑：王琦瑢
版式设计：王 晨　责任校对：周丽芳　责任印制：王 宏
印刷：中农印务有限公司
版次：2021年5月第1版
印次：2021年5月北京第1次印刷
发行：新华书店北京发行所
开本：889mm×1194mm 1/16
印张：9.25
字数：290千字
定价：228.00元

版权所有·侵权必究
凡购买本社图书，如有印装质量问题，我社负责调换。
服务电话：010-59195115　010-59194918